Human–Wildlife Conflict

DRILL HALL LIBRARY
MEDWAY

WITHDRAWN FROM UNIVERSITIES AT MEDWAY LIBRARY

SOS
1500
3014

4717671

Human–Wildlife Conflict

Complexity in the Marine Environment

Edited By

MEGAN M. DRAHEIM

Center for Leadership in Global Sustainability, Virginia Tech

FRANCINE MADDEN

Human–Wildlife Conflict Collaboration

JULIE-BETH McCARTHY

Independent Researcher

E. C. M. PARSONS

Department of Environmental Science and Policy, George Mason University

OXFORD
UNIVERSITY PRESS

Human–Wildlife Conflict: Complexity in the Marine Environment. Edited by Megan M. Draheim, Francine Madden, Julie-Beth McCarthy, and E. C. M. Parsons © Oxford University Press 2015. Published 2015 by Oxford University Press.

1 0 FEb 2016

OXFORD
UNIVERSITY PRESS

Great Clarendon Street, Oxford, OX2 6DP,
United Kingdom

Oxford University Press is a department of the University of Oxford.
It furthers the University's objective of excellence in research, scholarship,
and education by publishing worldwide. Oxford is a registered trade mark of
Oxford University Press in the UK and in certain other countries

© Oxford University Press 2015

The moral rights of the authors have been asserted

First Edition published in 2015
Impression: 1

All rights reserved. No part of this publication may be reproduced, stored in
a retrieval system, or transmitted, in any form or by any means, without the
prior permission in writing of Oxford University Press, or as expressly permitted
by law, by licence or under terms agreed with the appropriate reprographics
rights organization. Enquiries concerning reproduction outside the scope of the
above should be sent to the Rights Department, Oxford University Press, at the
address above

You must not circulate this work in any other form
and you must impose this same condition on any acquirer

Published in the United States of America by Oxford University Press
198 Madison Avenue, New York, NY 10016, United States of America

British Library Cataloguing in Publication Data
Data available

Library of Congress Control Number: 2015934243

ISBN 978-0-19-968714-5 (hbk.)
ISBN 978-0-19-968715-2 (pbk.)

Printed and bound by
CPI Group (UK) Ltd, Croydon, CR0 4YY

Links to third party websites are provided by Oxford in good faith and
for information only. Oxford disclaims any responsibility for the materials
contained in any third party website referenced in this work.

Preface

Conflict, or the potential for conflict, is inherent in human communities. The impacts of social conflict on conservation efforts are pervasive. Yet, conservation study and practice is still at a relatively early stage of understanding and addressing these impacts. Despite conservation's several-hundred-year-long history, it is only fairly recently that the field has broadened its disciplinary reach to include elements of psychology, anthropology, neurology, sociology, behavioral economics, systems thinking, and other human-oriented sciences into conservation research and practice. And in many ways we are still early in our journey to fully integrate the wisdom from these fields into what it means to do conservation.

Today many, perhaps even most, conservation researchers and practitioners intuitively understand the importance of conflict to their work, but typical educational and training paths do not develop the suite of skills and capacities needed to constructively transform conflict. In response, my organization—the Human–Wildlife Conflict Collaboration (HWCC)—adapted principles and approaches of conflict transformation developed over decades in the peace-building field, and introduced them to our field in 2008 as conservation conflict transformation. We continue to adapt, evolve, and improve our practice.

Recognition of the need to deepen our field's understanding of conflict was the inspiration and starting point for this book. This recognition is the first step toward transforming conflict so that it can support, rather than hinder, conservation. It is to the credit of this book's editors and authors that they took this goal to heart. Their willingness to learn this new approach and integrate conservation conflict transformation with their existing work and expertise demonstrates humility, courage, creativity, and adventurousness.

Our journey began in 2008, after Megan Draheim participated in a capacity-building workshop led by HWCC. A couple of years later, I was delighted when she invited me to co-edit a book that would interweave the conservation conflict transformation approach into a set of cases of marine-based human–wildlife conflict. We were subsequently enriched as an editorial team when Julie-Beth McCarthy and Chris Parsons joined us, bringing both marine conservation expertise and an open-minded willingness to learn about conservation conflict transformation. We sought, and found, chapter authors in both the research and practitioner communities who shared our sense of adventure and willingness to take on a new challenge.

For most of the contributors, this was their first encounter with conservation conflict transformation. A few had worked with me in one of HWCC's conservation conflict transformation capacity-building workshops in recent years. In every case, the editors and authors were enthusiastic to engage in short, intensive orientation or refresher seminars to build proficiency in one of the key analytical components of conservation conflict transformation—levels of conflict—as well as some of the principles of process design. They were then asked to apply that learning as part of the analysis in their case and chapter.

This was not an insignificant undertaking. Typically, an author in an edited volume is expected to write on a topic over which they have mastery. Rarely are authors (or editors) asked to go a step further, to learn and apply a new and very different approach and discuss their existing work or expertise in that new context. This required courage, patience, flexibility, and intellectual curiosity. It was an adventure in what futurist Alvin Toffler has said is the very definition of literacy in the twenty-first century: the capacity to learn, unlearn, and relearn. In this book, contributors are articulating not only what they know well but also new concepts they have begun to learn and apply within their areas of mastery.

In traveling this path, our team modeled the challenging stance that will be needed for conservation success: being willing to let go of what is known and comfortable and to remain open to and engaged with the paradoxical realities of a changing world. In doing so, our field will more readily embrace and more successfully engage with its increasing complexity to improve conservation outcomes.

Our goal in this book is to instill a sense of intellectual curiosity in you, the reader—the same curiosity that motivated us to embark on this expedition. As an exploration, this book does not offer solutions but rather insights and perspectives. This book is not prescriptive, nor should it be. Conflict and the specific processes needed to transform it are highly context specific. Having said that, we believe that the analysis of conservation conflict transformation across the wide range of marine settings discussed here argues persuasively for the broad applicability of conservation conflict transformation across a variety of cultures, contexts, species, and regions.

I should note that this book is intended to provide a small window into what it means to understand conflict through a conservation conflict transformation lens. By design, this book hones in on a single, important analytical tool—levels of conflict analysis—and demonstrates its replicability across many cases. That said, levels of conflict analysis is just one of several analytical tools in the conservation conflict transformation practitioner's toolbox. This toolbox also contains a suite of theories, principles, processes, strategies, and skills that are essential for transforming conflict but are largely beyond the scope of this book. While an understanding of levels of conflict analysis may give the reader (and the authors) deeper insights into cases, such insights do not necessarily translate into an immediate capacity to then transform the conflict they present. That capacity requires broader, deeper, and more holistic understanding of, personal engagement with, and years of practice in conservation conflict transformation. It is important to note that a complete engagement with conservation conflict transformation is outside the scope of this, or any, publication.

This book represents a stage in an intellectual expedition, in which a committed group of editors and authors explored the potential for a new perspective and analysis—that of conservation conflict transformation. Adopting a conservation conflict transformation perspective has enabled our authors to gain new and deeper insights into the social conflict and systemic complexity in their case studies in marine conservation. Conservation conflict transformation serves to compliment and augment the author's existing topical, disciplinary, species, and regional expertise. I admire and wish to thank the contributors for their curiosity, open-mindedness, and willingness to take risks and put in extra effort. I particularly appreciate authors such as Jill Lewandowski, Rachel Sprague, and Catherine Booker, who have been on a longer journey of understanding and integrating conservation conflict transformation within their work. Their efforts to provide leadership in their respective institutions, fields, and areas of expertise give me inspiration and great hope. I am proud of the work that our team has achieved in this book, and I hope that you find the reading of it to be as stimulating, informative, and thought-provoking as we found the writing.

Francine Madden
Human–Wildlife Conflict Collaboration
Washington, DC

Acknowledgments

Any book project is a large undertaking, especially one that was as much of a collaborative effort as this. As such, the editors, Megan M. Draheim, Francine Madden, Julie-Beth McCarthy, and E. C. M. (Chris) Parsons, have a lot of people to acknowledge. Most importantly, we'd like to send our sincere thanks to our authors for all of the time and effort they put into this project to make it a success. They brought their own extensive experience in a wide variety of fields to the table, were all willing to tackle what was in most cases a new paradigm, and shared their knowledge and expertise as a reviewer for their co-contributors as well. We cannot thank them enough for their energy, time, and patience throughout. We would also like to thank Lucy Nash and Ian Sherman at Oxford University Press for all of their help and support throughout this process. They were both a delight to work with.

Megan M. Draheim would first like to thank her co-editors for being willing to start this project in the first place and for all of their work throughout. Her co-author on Chapter 7, Rachel Sprague, deserves special thanks as well. Kieran Lindsay and Heather Eves provided invaluable feedback and support, especially as the project neared completion, and she hopes to be able to return the favor in the future. Her parents, James Draheim and Linda Jo Clough, her stepfather, Rodney Clough, and countless friends have helped in many ways, big and small. Her husband, David Harris, deserves special credit for tirelessly letting her discuss the project and for his helpful comments throughout this process. For that, and for so much more, she can never thank him enough.

Francine Madden is grateful to her co-editors and all the authors for their enthusiasm for learning about and addressing levels of conflict and conservation conflict transformation in their chapters, as well as their boundless patience through the editing process. Words can't express the gratitude Francine has for her husband, partner, and best friend, David Downes; among other virtues, he is a gifted editor. Francine is appreciative of and inspired by her team at the Human–Wildlife Conflict Collaboration (HWCC), and all the wonderful leaders, practitioners, and stakeholders she has had the honor and pleasure of working with these last nine years. And to Francine's co-author on Chapter 1, Brian McQuinn—HWCC would not be where it is today without his critical support, generosity, and intelligence during those early years.

Julie-Beth McCarthy would first like to thank her co-editors for bringing her on board and engaging in such a unique and interdisciplinary project. Kendra Marks must be thanked profusely for bravely diving into the abyss, armed with her incredible research skills, and providing invaluable assistance early on. Julie-Beth would also like to thank her parents (Paul McCarthy and Maureen Duke-Renouf) who have long supported her various endeavors and who helped out in their own ways from both near and far. Finally, though most importantly, she would like to thank her husband, Ian Sturgess, for his patience, advice, and unwavering support; and her kids, Solveig and Healy, for ensuring that things are always kept in perspective.

Chris Parsons thanks Naomi Rose for her careful proofreading, correction of his dyslexic lapses, and editorial advice; he also thanks various delegates at the IWC, from both sides of the argument, who engaged in numerous discussions about the whaling issue in a number of bars and pubs around the world.

Contents

List of Contributors

Catherine Booker, Community Conch, George Town, Exuma, Bahamas

Megan M. Draheim, Center for Leadership in Global Sustainability, Virginia Tech, Arlington, VA, USA

Christine Gleason, Department of Environmental Science and Policy, George Mason University, Fairfax, VA, USA

Jill Lewandowski, Department of Environmental Science and Policy, George Mason University, Fairfax, VA, USA, and Bureau of Ocean Energy Management, Herndon, VA, USA

Francine Madden, Human–Wildlife Conflict Collaboration, Washington, DC, USA

d'Shan Maycock, EConnect, Freeport, Grand Bahama, Bahamas

Julie-Beth McCarthy, Independent researcher, Sparwood, BC, Canada

Brian McQuinn, St Cross College, University of Oxford, UK

E. C. M. Parsons, Department of Environmental Science and Policy, George Mason University, Fairfax, VA, USA

Rachel S. Sprague, Pacific Islands Regional Office, NOAA National Marine Fisheries Service, Honolulu, HI, USA

Carlie Wiener, Faculty of Environmental Studies, York University, Toronto, ON, Canada

Sarah Wise, MARUM, Center for Marine Environmental Sciences, and ARTEC, Institute for Sustainability Studies, University of Bremen, Germany

List of Acronyms

ATOC: Acoustic Thermometry of Ocean Climate
BEST: Bahamas Environment Science and Technology Commission
BNT: Bahamas National Trust
CEBSE: Conservación y Ecodesarrollo de la Bahía de Samaná y su Entorno
DMR: Department of Marine Resources
EHMSC: Elizabeth Harbour Management Steering Committee
FRIENDS: Friends of the Environment
GEF: Global Environment Facility
HWCC: Human-Wildlife Conflict Collaboration
IWCAM: Integrating Watershed and Coastal Area Management
IWC: International Whaling Commission
LFA: low-frequency active sonar systems
MFA: mid-frequency active sonar systems
MPA: marine protected area
NMFS: National Marine Fisheries Service
NOAA: National Oceanic and Atmospheric Administration
NSF: National Science Foundation
PG&E: Pacific Gas and Electric
REA: Rapid Ecological Assessment
RMP: Revised Management Procedure
SIDS: Small Island Developing States
SNS: sacred natural sites
SPZ: seal protection zone
TEK: traditional ecological knowledge
UNCLOS III: Third United Nations Conference on the Law of the Sea
WNP: West Side National Park

Introduction

While human–wildlife conflict has long been recognized as a serious conservation threat within the wildlife conservation community, there have been too few opportunities for the sharing of lessons learned and communicating best practices in understanding and addressing the social complexity within the human–wildlife conflict context, perhaps especially in the marine realm. Human–wildlife conflict has classically been defined as a situation where wildlife impacts humans negatively (physically, economically, or psychologically), and where humans likewise negatively impact wildlife. However, there is growing consensus in the human–wildlife conflict community that the conflict between people about wildlife is as much a part of human–wildlife conflict as is the conflict between people and wildlife. Human–wildlife conflict not only affects the conservation of one species in a certain geographic area but also impacts an individual's, community's, and society's desire to support conservation programs in general.

Human–Wildlife Conflict: Complexity in the Marine Environment explores the complexity inherent in situations where human–wildlife conflict plays a role in influencing human actions. The book covers the theory, principles, and practical applications of human–wildlife conflict work, making it accessible and usable for conservation practitioners, as well as of interest to researchers more concerned with a theoretical approach to the subject.

This book is the culmination of several years of work by our entire team. As Francine Madden mentioned in the preface to this book, this was an ambitious project—contributors were expected to not only write about their field of expertise, but they were also given the task of incorporating an innovative theoretical framework, conservation conflict transformation, and more specifically a conservation conflict transformation analysis tool, the levels of conflict model, which was in most cases a new mode of thinking about their case studies. This was challenging, and they deserve great credit for their efforts. From the beginning, we wanted to explore the connections between diverse marine conservation conflict cases. The levels of conflict model gave us a framework in which to do so. Our ultimate goal was to paint a rather broad canvas of case studies that demonstrates the complexity of human–wildlife conflict in the marine environment. As such, we were hesitant to break the chapters into what are ultimately rather arbitrary sections. However, providing some structure to the book was necessary, so three sections were created for the sake of clarity: "Introduction to the levels of conflict," "Policy and human–wildlife conflict," and "Narratives and human–wildlife conflict." Each chapter includes both the thematic levels of conflict analysis and a discussion about the authors' cases from a different disciplinary lens, ranging from the role that storytelling can play in conflict to the challenges of wicked problems and common pool resources, among others.

We wanted to ensure that the connections between all the chapters and sections were clear. While including a levels of conflict analysis in each chapter provided a strong connection across the entire book, we also drew links across the chapters and sections in order to underline this point. Our hope is that it will show the reader that similar elements are found in most case studies of human–wildlife conflict. To further make the book a cohesive whole, the editors also created a short "lessons learned" textbox at the end of each chapter. These are not the only lessons to be had from these rich chapters; rather, they are points that stood out to us as editors.

Although our case studies are marine oriented, our strong belief is that conservation researchers and practitioners who work in terrestrial systems could also benefit from the case studies in this book. At the outset of our project, one of our stated goals was to bring this work to the marine conservation community, but we believe that there are lessons to be learned from these case studies across the entire conservation community. Therefore, our hope is that the audience for this book will not be limited to marine conservationists but will include conservationists across all disciplines.

Our first section, "Introduction to the levels of conflict," has a standalone chapter. Here, Francine Madden and Brian McQuinn provide an introduction to conservation conflict transformation and the levels of conflict model in "Understanding social conflict and complexity in marine conservation." Madden and McQuinn provide an overview of conservation conflict transformation, an innovative framework for understanding and dealing with conservation-related conflict. The levels of conflict model is a conflict analysis tool in the conservation conflict transformation suite and is used throughout this book. Madden and McQuinn provide the introduction to this powerful instrument, as well as providing insight into how to tackle the complex conflict that conservation practitioners and researchers are often faced with.

Our second section, "Policy and human–wildlife conflict," contains five chapters. Catherine Booker and d'Shan Maycock discuss two conservation interventions in the Bahamas in their chapter "Conservation on island time: stakeholder participation and conflict in marine resource management." The first case involves a community-based harbor management initiative, and the second, a program to help make the country's spiny lobster fishery more sustainable. Booker and Maycock explore the different approaches that were brought to these projects, and why one was ultimately more successful than the other. They close with a discussion of best practices when it comes to participatory processes that must tackle conservation-related conflict.

Jill Lewandowski tackles wicked environmental problems in "Transforming wicked environmental problems in the government arena—a case study of the effects of marine sound on marine mammals." Anthropogenic noise in the marine environment has been an ongoing controversy in the United States, involving everyone from environmental groups to the U.S. Navy, to energy and other business interests. It also meets all of the standards of a wicked environmental problem (it has scientific uncertainty, political and regulatory complexity, competing stakeholder interests, a background of conflict and distrust between parties, and decision-making processes that only lead to further conflict). Lewandowski explores anthropogenic noise as a means to not only discuss wicked environmental problems and provide a levels of conflict analysis but also to discuss ways in which regulatory agencies might improve their decision-making processes to transform these conflicts and even prevent wicked environmental problems in the first place.

Christine Gleason's chapter, "Conservation in conflict: an overview of humpback whale management in Samaná, Dominican Republic," offers a look at a co-management program for whale-watching, a common pooled resource of great economic import in the areas around Samaná Bay, Dominican Republic. Whale-watching is increasingly big business across the world, and regulations vary widely from place to place. After an initial attempt at voluntary regulations spearheaded by two NGOs failed, a co-management system was created in 1998. However, conflicts between those who are on the inside of the system and those who are not have plagued these regulatory attempts. Some of these conflicts reflect not only present-day economic concerns but also the complex colonial and racial history of the Dominican Republic.

E. C. M. Parsons offers his experience as a scientific delegate to the International Whaling Commission (IWC) in "Levels of marine human-wildlife conflict: a whaling case study" to share some insight into how and why conflict has been so pervasive in that organization in recent decades, as well as looking forward to some strategies that might prove successful in transforming that conflict. Much of the conflict within the IWC lies in the fact that Japan, a signatory to the Convention, engages in whaling through a clause in the treaty that allows whales to be killed for scientific research purposes. Parsons discusses why this conflict goes well beyond a disagreement over what might constitute valid scientific research and instead is embedded in questions over cultural differences between the countries at opposite ends of the conflict.

We return to the Bahamas, and in particular to Andros Island, in Sarah Wise's "Conflict and collaboration in marine conservation work: transcending boundaries and encountering flamingos." Wise's chapter describes what happened when a new marine protected area (MPA) was proposed on the western side of Andros. The proposal was met with criticism by many locals, who believed the new park would disproportionately benefit wealthy residents while hurting those who rely on the area's natural resources. This belief was predicated on both recent (including the establishment of a different MPA) and past history (including a legacy of colonialism and racial injustice) and endangered the effectiveness of the MPA.

Our third section, "Narratives and human–wildlife conflict," has three chapters that focus on how narratives can play a role in either advancing conflict or advancing conflict transformation. Rachel S. Sprague and Megan M. Draheim write about Hawaiian monk seals in "Hawaiian monk seals: labels, names, and stories in conflict." They discuss how the recent increase in monk seals in the human-inhabited main Hawaiian Islands has created conflict between those who believe monk seals "belong" to the islands and those who do not. Stories about monk seals—and the lack thereof—have informed this debate, and the seals, in some cases, are a proxy for deeper conflict over who "belongs" to the islands and who does not, speaking to native Hawaiian rights and the state's history of colonialism.

Dolphins have been a potent symbol in human culture and discourse for thousands of years, and Carlie Wiener takes a look at how this can impact present-day conflict over dolphins as it relates to "swim-with-dolphin" programs in Hawaii in "Flipper fallout: dolphins as cultural workers and the human conflicts that ensue." This conflict is much more complex than simply those who are in favor of swimming with dolphins against those who are not, however. These two populations are further split along other (sometimes overlapping) lines, including those who make a living from swimming with dolphin trips, those who do not, those who say they have a spiritual connection with the dolphins, those who believe the government should or should not regulate such interactions, and native Hawaiians. Weiner also provides an overview of cultural representations of dolphins in media such as movies, books, and television shows and discusses how these can help advance conflict between people about dolphins.

And finally, Julie-Beth McCarthy, in "Examining identity-level conflict: the role of religion," delves into the role that religion can play in human–wildlife conflicts, as both a positive and a negative force for conservation. Religions have a long history with conservation, both purposeful and incidental, through teachings, sacred natural sites, traditional ecological knowledge, and taboos. She explores two case studies: one where religion was successfully used to help transform conflict around whale shark hunting, and one where religion has acted as a stumbling block for the conservation of an MPA. In both cases, she shares strategies that conservation professionals might use when navigating the religious side of human–wildlife conflict.

Megan M. Draheim,
Francine Madden,
Julie-Beth McCarthy, and
E. C. M. Parsons

Section 1

INTRODUCTION TO THE LEVELS OF CONFLICT

1

Understanding Social Conflict and Complexity in Marine Conservation

Francine Madden and Brian McQuinn

1.1 Introduction to conservation conflict transformation

The disagreements and seemingly incompatible agendas that are commonly found among groups and individuals within marine conservation contexts can deepen into intractable conflict because of deeper social conflicts that may have little to do with the stated dispute (Coleman, 2011; Madden and McQuinn, 2014). Human–wildlife conflict and other conservation conflicts are, indeed, challenging obstacles for conservation globally (Madden, 2004; Michalski et al., 2006; Peterson et al., 2013; Redpath et al., 2013). To enable the conditions for durable, mutually supported decision-making, changes in process and improvements to key relationships are paramount. Yet, while there is growing recognition in conservation for the need to advance collaborative governance models, too often the individuals or organizations driving the process fail to recognize the conflict dynamics or support appropriate processes that reconcile the deep-rooted conflict among stakeholders (Madden and McQuinn, 2014). As a result, even when problems are solved, these solutions either do not last, or when they do, they fail to create the desired broader change needed to support long-term conservation efforts (Rothman, 1997; Balint et al., 2011; Doucey, 2011; Peterson et al., 2013; Madden and McQuinn, 2014).

As a first step in addressing these deeper conflicts that undermine conservation goals, analysis of conflict dynamics is critical (Madden and McQuinn, 2014). Conflict analysis is, at the very least, an essential "do no harm" step to reduce the risk that project decision-making and implementation inadvertently create or exacerbate conflict. Appropriate analysis will also give the practitioner strategic insights into how to design decision-making processes so that the decisions, solutions, and projects are supported by a mutually respectful, resilient, engaged community of stakeholders.

Current analyses, when they do focus on the human dimension, typically focus on values, opinions, and attitudes toward the wildlife or toward management actions. They often

Human–Wildlife Conflict: Complexity in the Marine Environment. Edited by Megan M. Draheim, Francine Madden, Julie-Beth McCarthy, and E. C. M. Parsons © Oxford University Press 2015. Published 2015 by Oxford University Press.

emphasize the behavior of the wildlife or the material and economic consequences for people. Social conflict is often manifested in people's reactions to and retaliations against the wildlife that governments and NGOs are seeking to conserve (see Chapter 7; Dickman, 2010), but conservation analysis rarely takes adequate account of the deeper social conflict dynamics between the people and groups often entangled in these disputes (Jeong, 2008; Coleman, 2011; Deutsch and Coleman, 2012; Peterson et al., 2013). As a result, stakeholder engagement processes often overlook this hidden dimension of conflict that, if accounted for, would help foster social receptivity to change and create the conditions for more sustainable long-term agreements (Lederach, 1997; Rothman, 1997; Jeong, 2008; Levinger, 2013). There is an inclination toward negotiation and other superficial, issue-oriented processes that further ignores these complex social conflicts (Fisher et al., 1991; Leong et al., 2009; Dickman, 2010; Balint et al., 2011; Coleman, 2011; Doucey, 2011). Too often, a limited capacity to understand the social conflict dynamics, a limited willingness, capacity, or authorization to alter process design, and anxiety about engaging with the intricacies of human interactions and needs all conspire to hinder the success and durability of conservation actions (Ansell and Gash, 2008; DeCaro and Stokes, 2008; Manolis et al., 2009; Coleman, 2011; Leong et al., 2011). Such omissions can inflame the deeper social and psychological drivers of complex conflict, of which the surface manifestations, such as people's resentment of wildlife, are only one small dimension (Burton, 1990; Rothman, 1997; Lederach, 2003; Dickman, 2010; Madden and McQuinn, 2014).

As illustrated by the cases discussed in this volume, durable success in marine conservation necessitates a deepening of conservation and stakeholder capacity and a broadening of approaches to address these deeper drivers of conflict (Madden, 2004; Deutsch et al., 2006; Manolis et al., 2009; Dickman, 2010; Peterson et al., 2013; Madden and McQuinn, 2014). Conservation conflict transformation provides the base needed to reorient the conservation field's capacity and strategies in this regard, through adaptation and augmentation of the conflict transformation subdiscipline within the peace-building sector. Using a suite of theories, principles, processes, and strategies that are a better match to the needs and conditions of conflict in conservation (Madden and McQuinn, 2014), conservation conflict transformation aims to constructively transform destructive social conflicts by targeting change in the quality of relationships and the process structures and systems underpinning and undermining conservation efforts.

This chapter outlines the current limitations in prevailing approaches and capacities in addressing marine-based human–wildlife conflict and related conservation conflict; defines conservation conflict transformation as a perspective, an approach, and a goal; summarizes a model for conflict analysis called levels of conflict (applied throughout this book); and provides an overview of the essential elements for conflict intervention.

1.2 The conceptual and capacity limitations of current conflict approaches

Conservation's attention to the human side of human–wildlife conflict or other conservation conflicts is typically too narrow to be effective. While the field is evolving, conservation efforts still tend to be focused on physical and spatial measures (e.g., zoning beaches or bodies of water), economic fixes (e.g., providing jobs or offering incentive or compensatory payments), technical solutions (e.g., changes to fishing or boating practices), legal actions (e.g., laws and law enforcement that are designed to protect wildlife and their habitat), and biological methods (e.g., assessing population viability with respect to harvest levels). While these considerations are needed for the success of conservation, we suggest they are inadequate when taken alone without also addressing the social and psychological needs that drive social conflict (Dukes, 1999; Lederach, 2003; Blackstock et al., 2007; Reed, 2008; Leong et al., 2009; Balint et al., 2011; Leong et al., 2011; Peterson et al., 2013; Madden and McQuinn, 2014). Certainly, many conservation professionals advocate for greater stakeholder engagement in conservation decisions (Treves et al., 2009; Barlow et al., 2010; Redpath et al., 2013). Yet, capacity to design and lead transformative processes that target the social, systemic, and psychological drivers of conflict, improve trust, and increase stakeholder receptivity to shared goals remains deficient across the conservation field (Reed, 2008; Leong et al., 2009; Manolis et al., 2009; Leong et al., 2011; Madden and McQuinn, 2014).   → *Address inherent dynamics.*

The more widely accepted current approaches to human–wildlife conflict rely on an assumption that the economic or physical costs associated with human–wildlife conflict are the most pressing aspects of the conflict that need resolution for coexistence to exist. This assumption relies, implicitly, on Abraham Maslow's "hierarchy of needs" (Maslow, 1954). Maslow's theory suggests that the most fundamental human needs are physiological (food, water, shelter, and sleep) and security (physical, financial, health, and property). Maslow hypothesized that until what he calls the "basic" needs are met, humans are less concerned with or do not seek out the "higher-level" social and psychological needs. These social and psychological needs include respect and recognition; identity, or how one sees themselves or their group in relation to the rest of the world; social, emotional, or cultural security; freedom of movement and in making decisions; connectedness and belonging; meaning in terms of place or context; and an ability to reach one's potential (Burton, 1990; Marker, 2003).

However, Maslow's framework has been repeatedly refuted by scholars from a variety of disciplines and fields, including sociology, psychology, peace-building, and economics (Coate and Rosati, 1988; Max-Neef et al., 1989; Burton, 1990; Clark, 1990; Galtung, 1990). These scholars have argued that people tend to pursue all these needs simultaneously and doggedly (Marker, 2003). They are, in fact, not sought out in a hierarchical order or stepwise progression.

According to Burton, social and psychological needs, when unmet, serve to drive conflict toward intractability (Burton, 1990). When unmet, these needs also explain the existence of what would otherwise seem to be "irrational" or self-sabotaging reactions to attempts to resolve conflict. Many conflicts in conservation deepen and persist because these needs are ignored or inadequately addressed. For example, a conservation authority's presence, programs, and resources devoted to marine wildlife needs may unintentionally convey a message that the wildlife is more important than the people, leaving people feeling disrespected and disenfranchised (Madden, 2004).

The influence of these deeper social and psychological needs on conservation effectiveness is unmistakable, even if it is not always obvious. Conservationists themselves rarely examine how their own values and beliefs shape their worldview, drive their sense of moral superiority, or are perceived (either implicitly or explicitly) by others in conflict settings (Pearce and Littejohn, 1997; Madden and McQuinn, 2014). Similarly, other stakeholders in conservation conflicts fail to fully understand or describe the needs, values, and beliefs driving their behavior in ways that can foster meaningful and constructive inclusion of their needs in conservation decisions. In one example, a community resource user initially articulated that financial loss was his predominant grievance regarding human–wildlife conflict and with respect to the government and conservation NGO authorities. He reiterated that the financial hardship associated with wildlife was the driving factor in his resentment. Yet, this same resource user later noted that he refused substantial government or conservation financial support because, to him, accepting that money was akin to accepting the wildlife and forging an alliance with the government. Doing so would be a direct threat to his way of life and it would be traitorous to his group of resource users. This was, in fact, a more profound determiner of his behavior than the financial impact.

Indeed, complaints regarding the fairness of material or resource disputes often serve as an opportunity to right perceived wrongs from the past, continue or escalate the fight to protect one's individual or group identity, or balance the power in asymmetrical relationships (Burton, 1984). The practical result for conservationists is that if these deeper conflict dynamics are ignored, any solution to what appears to be the problem will be, at best, a temporary fix (Rothman, 1997). Yet, too often, efforts to engage stakeholders in conservation-oriented decision-making focus superficially on the easily seen or acknowledged aspects of a conflict. Analysis often ignores the complex conflict dynamics, and decision-making processes are not designed to engage with, untangle, or reconcile the deeper drivers of the conflict. As a result, even if a tangible problem is "solved," no real change is generated.

Regarding conservation's capacity to engage with social conflict, we identify additional limitations. Typically, wildlife professionals enter the field because of an interest in conserving wildlife and other natural resources—not working with people, especially people who might be hostile to conservation. While educational and professional development opportunities in the "human dimensions" of fish and wildlife management have dramatically increased in recent decades (Conover, 2001; Decker et al., 2012), professionals with exposure to this subfield still often lack practical "know-how" for engaging in conflict constructively, analyzing complex social conflict, or designing effective interventions to address it successfully and sustainably (Decker et al., 2012; Madden and McQuinn, 2014). Indeed, most wildlife professionals have not been provided with the practical training and skills needed to work effectively to reconcile intractable conflicts, develop productive and sustainable collaborations that promote systemic change, develop and build on common ground with people and groups who may be deeply hostile to conservation and not share these values, or collaborate with groups and individuals whose social identity may be threatened by wildlife authority decisions, or indeed the very existence of such decisions, and who may feel disempowered or angry about past or present decisions (Madden, 2004; Lederach et al., 2007; Madden and McQuinn, 2014).

Even where curricula include stakeholder engagement and conflict management, they often solely focus on "getting to common interests" that may be inaccessible or unable to be realized if deeper-rooted social conflict is at play (Lederach et al., 2007). The need for public participation, community involvement, or transparent decision-making are often acknowledged (Conover, 2001; Reed, 2008; Treves et al., 2009; Decker et al., 2012; Redpath et al., 2013), but training (and implementation) is usually oriented toward negotiation, arms-length notice-and-comment procedures, public "town hall" meetings, linear multistakeholder processes that overemphasize decision-making and underappreciate the social and psychological changes that need to take place before decisions can be genuinely agreed upon and supported, and efforts to simplify and assimilate competing values and scientific uncertainty into quantifiable variables (Chapter 3; Madden and McQuinn, 2014). These approaches may be useful in some contexts, but they are often insufficient for social conflicts typical of wildlife conservation contexts.

Despite these gaps, some professionals find effective ways of untangling conflict at deeper levels in the field, but they often work instinctively and over years of trial and error. Such an arduous and time-consuming process wastes limited resources. What's more, typically these professionals are at a loss to identify all the relevant factors, and thus they and others are hindered from reproducing such achievements in other efforts. This deficiency in explicit and widespread knowledge and application in the field seriously hinders conservation's systematic progress globally. If integration is held to a high standard of practice, conservation conflict transformation offers an opportunity to overcome these field-wide inefficiencies and promote greater effectiveness not just for the sake of conservation, but also for the well-being of society.

1.3 Conservation conflict transformation

1.3.1 What is conservation conflict transformation?

Conservation conflict transformation is a perspective, an approach, and a goal (Madden and McQuinn, 2014). Conservation conflict transformation as a perspective recognizes that conflict is a fundamental part of human communities and human interaction. Conflict is not inherently bad; it may be deeply destructive, but can also be an opportunity for constructive change. Conservation conflict transformation also recognizes that the energy bound up in destructive conflict can be transformed into a constructive, creative, and collaborative energy if decision-making processes are designed appropriately and fractured relationships are reconciled (through an appropriately designed and facilitated process). Conservation conflict transformation also recognizes that the interplay of conflict occurs at different scales, from the interpersonal to the multistakeholder, and as part of complex social systems. Transforming conflict from destructive to constructive is possible and requires strategic and creative engagement at and between all scales.

Conservation conflict transformation as an approach stems from a distinct theory and method that developed out of a reconceptualization of conventional theories and approaches aimed at ensuring more appropriate responses to contemporary conflicts (Miall, 2004). Conservation conflicts are often entrenched, prolonged, complex, and interrelated at micro and macro scales. Conservation conflict transformation fosters reconciliation of destructive relationships, and engagement of diverse and often marginalized participants. It engages with the complex nature of conflict and seeks long-term processes to ensure resilience within the system. Conservation conflicts include differing and deeply held values and identities, high emotions and high stakes, power imbalances, and a sense of moral authority that may push parties to maintain the fight, even when winning in the short term is not likely (Pearce and Littlejohn, 1997; Clark, 2002; Burgess, 2004). Nonnegotiable social and psychological needs are often at the base of conflicts that may present themselves as involving negotiable, tangible needs (Burton, 1990, 1993). When threatened, identity needs, especially, generate substantial destructive reactions (Rothman, 1997). Deep-rooted conflicts typically have conflict both within groups (intragroup) and between groups (intergroup), where intragroup conflict actually perpetuates intergroup conflict in an effort to protect identity and ensure group cohesion (Deutsch, 1973; Deutsch and Coleman, 2012).

Conservation conflicts are often impacted today by the antagonistic history of events, decisions, and relationships. This history contributes significantly to the current feelings of hostility, futility, and seeming intractability, making the current disputes more and more difficult to resolve. It often seems as if there is no way out or that the current state of affairs is simply just how things will always be. These views themselves can further perpetuate destructive action in a vicious cycle of conflict (Deutsch et al., 2006; Deutsch and Coleman, 2012). Ironically, such conflicts often lead disputants to harm themselves, and that which they value, in an attempt to make certain their adversary does not win (Atran and Axelrod, 2008).

Ensuring a comprehensive analysis of all levels of conflict is a critical first step in any conservation conflict transformation approach (Madden and McQuinn, 2014). At a minimum, conservation conflict transformation analysis supports efforts to "do no harm" when intervening in a conflict system. This is important because often conservation actions either cause unintended negative consequences or miss critical opportunities due to a lack of understanding of the conflict dynamics within the system in which conservation actions take place. There are multiple analytical tools a conservation conflict transformation practitioner uses to comprehensively understand complex conflict dynamics. That said, within the scope of this book, the authors largely explore one essential tool, levels of conflict, through a variety of cases in marine-based conflict.

The goal of conservation conflict transformation is to address the immediate, presenting issues while also reconciling destructive relationships and positively reforming the underlying processes, structures, and patterns that are the driving forces in cycles of destructive conflict. Conservation conflict transformation processes are a dynamic convergence of the relational context, positive support for constructive change processes, and the recognition that conflict is an opportunity to be embraced, not attacked or avoided (Lederach, 2003).

1.3.2 Levels of conflict: an essential tool for analysis

The levels of conflict model is an analytical tool for exploring the types and intensity of conflict that may exist in a conservation conflict context (Madden and McQuinn, 2014). An exploration of conflict using this model is a critical starting point for determining how one should engage with the conflict and seek its transformation. The model, initially developed by the Canadian Institute for Conflict Resolution, categorizes three levels of conflict: disputes, underlying, and identity based (Canadian Institute for Conflict Resolution, 2000; Madden and McQuinn, 2014; Figure 1).

The dispute is the first level of conflict. Disputes are the observable, palpable, immediate expression of a conflict (see Figure 1). People tend to feel comfortable asserting their issues and needs about a dispute, but in doing so they often create the illusion that this is their main or only source of distress in the conflict. (Sometimes that is the case, as disputes can indeed exist on their own; but they rarely do.) For instance, a dispute could center on a frustration that monk seals are eating a fisher's fish (see Chapter 7), whether or not data from Japan's scientific whaling program are valid (see Chapter 5), or whether pollution in Elizabeth Harbor, Great Exuma, stems mostly from boats or from onshore sources (see Chapter 2). If all that was going on in these cases was the dispute, it would be relatively easy to settle these issues. Yet, a limited focus on the "dispute" level, if there are other levels of conflict at play, may partly explain why even when the dispute is settled, the conflict persists.

The next deeper level of conflict that may exist is underlying conflict. Underlying conflict is a history of unresolved disputes. Underlying conflict exists when past interactions, decisions, or events leave one or both parties feeling dissatisfied, disrespected, or disempowered. When underlying conflict exists, it will saturate any presenting dispute with added meaning, emotion, and weight that may seem disproportionate to the concerns at stake in the dispute alone. The existence and significance of underlying conflict may be masked by the disputants themselves, as it is often easier to attend to and talk about the more concrete and specific issues involved in the current dispute than it is to express the feelings about how previous decisions were made.

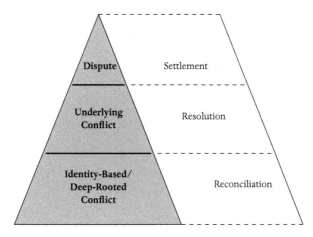

Figure 1.1 The three levels of conflict that may exist in any conservation conflict context (and the corresponding process needed to address conflict at that level). Source: Madden and McQuinn, 2014 reproduced with permission from Elsevier.

Given the long-term engagement and history of conservation efforts, it is virtually impossible to escape the existence of underlying conflict in any given context. The following examples demonstrate some of the diversity of examples of underlying conflict: "swim-with-wild-dolphins" program operators in Hawai'i being upset with the federal government about repeated attempts over a long period of time to regulate the industry, leading to uncertainty about their businesses' futures (see Chapter 8); disagreement over the way that land was selected for multiple protected areas (see Chapter 6); and a history of extensive litigation between the U.S. Navy and some NGOs over the Navy's use of sonar (see Chapter 3).

Identity-based, or deep-rooted, conflict is the third level of conflict. Identity-based conflict involves deeply held values, beliefs, or social–psychological needs that are fundamental to the identity of at least one of the parties involved in the conflict. When identity conflict is present, individuals and groups make prejudicial assumptions and judgments about others based on their group affiliation. They may assign responsibility to the other individual for past actions taken by other members of their group. This added level of conflict drives conflict toward seeming intractability as it contributes more severity and complexity to the current situation. This distorted lens of prejudice impacts an individual's or group's ability to work effectively toward a shared decision because the very act of working together may be a threat to the identity of one or more groups.

Many conservation conflicts involve conditions where one group's identity may actually be defined in opposition to another's because of perceived threats to their identity or way of life. Identity conflict exists both between groups and within groups. Even conflicts that begin as a dispute may evolve into an identity conflict over time as those involved in the dispute come to identify themselves and their group with a particular position in the dispute. Eventually, these issues become so entrenched that they become a core part of the identity of the individual or group. When identity conflict is at play, actions and reactions may appear "irrational" to the uninformed observer or participant.

As diverse as human nature is, so are the potential expressions of identity conflict. In marine-based conservation, identity conflict may exist where MPA managers believe that members of a particular indigenous group are unreliable and untrustworthy partners in conservation because of their spiritual beliefs and indigenous affiliation (see Chapter 9); where large whale-watching boat operators instinctively ignore the experience and knowledge of their captains and crews because of differing socioeconomic status (see Chapter 4); and where native Hawaiians retaliate against Hawaiian monk seal conservation efforts because of the monk seal's association with "outsiders" (e.g., nonnative Hawaiians, people from the mainland, and, in particular, the U.S. government) who are perceived to be a threat to the Native Hawaiian identity and way of life (see Chapter 7).

While disputes tend to be concrete, measurable, and easily discernible, underlying and identity-based conflicts are often elusive, difficult to quantify, and either poorly expressed or challenging to address. An underlying and identity-based conflict may be manifested as a dispute in order to give an easier, more palpable, and clearer focus to a group's distress (Rothman, 1997). Moreover, articulating emotional or psychological needs or

injuries is not easy, particularly in contexts where the level of trust is low (Sites, 1990). Finally, conservation's focus on wildlife may implicitly narrow the parameters of what is acceptable for discussion so as to discourage disclosure of the impact of conservation decisions and actions on an individual's social and psychological well-being (Clark, 2002; Madden and McQuinn, 2014).

Different processes are needed to address conflict at the different levels. The deeper the conflict, the more involved the process. If all that exists is a dispute, the process—settlement—is fairly simple. Disputes in many societies are commonly settled in courts using a rights-based system with legal codes for assigning responsibility, securing evidence, and determining outcomes. The use of lawsuits, laws, and law enforcement address conflict at a dispute level. Compliance with a "settlement" may address the immediate dispute, but if deeper levels of conflict exist and are not addressed appropriately, settlements are often temporary or inadequate, and those involved will likely use (or create) another occasion to rectify alleged injustices (Madden and McQuinn, 2014).

The term "resolution" is used to describe processes designed to address underlying conflicts, while the term "reconciliation" suggests the more involved process and profound identity shifts needed to address deep-rooted conflicts (Madden and McQuinn, 2014). Employing a dispute-level settlement process when deeper conflicts are at play may temporarily solve a problem, but will not create the kind of positive, sustained change conservation seeks. In fact, Ginges et al. (2007) found that when identity-based conflict is evident, the intensity of the conflict can actually grow rather than shrink when the deeper-rooted conflicts are ignored and material incentives (dispute-level tactics) are proposed as a concession.

Successful outcomes in conservation contexts are often hindered because of a prevalence of underlying and identity conflict in conservation coupled with an overuse of stakeholder engagement processes that are designed largely for dispute-level conflicts (and sometimes a partial treatment of underlying conflict). By increasing conservation's capacity to understand and address the complex social conflict dynamics and simultaneously recognize the systemic relevance of these conflicts for conservation (even as they may seem unrelated to conservation or have their roots elsewhere in society), conservation may overcome one of the greatest threats to its success (Madden and McQuinn, 2014).

1.3.3 Critical aspects of intervention

Conflict can arise from three sources: the substance of an issue or decision, the process for making those decisions, and the relationships among those involved (Madden and McQuinn, 2014). As discussed, there is often an overemphasis on the dispute or the manifest substance of the conflict. However, while all three aspects of conflict are important, the process and relationship dimensions of a conflict intervention offer a critical opportunity to address underlying and identity-based conflicts. Therefore, an effective conflict intervention will ensure equal attention to all three.

Process factors relate to decision-making design, fairness, control or power, and how (and by whom) these are employed. Parties might realize a specific solution is in their best interest, but if they do not feel the process gave them voice and was legitimate and fair, they may reject the solution regardless. Equally, parties will more readily accept and support decisions whose substance is less than ideal from their perspective, if they felt genuinely respected and empowered in the decision-making process (Fisher et al., 1991; Reed, 2008; Leong et al., 2009).

Lovallo and Sibony (2010) found that the quality of the decision-making process was six times more influential in determining the success and sustainability of a business decision than was the analysis of the potential solutions. Well-designed and facilitated decision-making processes improve relationships, which further support creativity and durability of solutions. Constructive respectful relationships mean communication and trust are also improved, and this leads to a greater capacity for resilience and adaptability as new challenges inevitably arise (Ansell and Gash, 2008; Reed, 2008).

Institutions and individuals with greater formal power and control often resist efforts to endorse genuinely equitable decision-making processes because they fear a loss of their authority or that an unacceptable compromise will result. Yet, anecdotal evidence from conservation and from research in other sectors contradicts these preconceptions. Such processes can actually lead to better decisions and an expanded range of mutually beneficial results when processes address not just the dispute at hand but also the unmet human needs that drive underlying and identity conflict (Lederach, 1997, 2003; Anderson and Olson, 2003; Lederach et al., 2007; Hendrick, 2009).

Stakeholders in a decision-making process often undervalue or undermine decisions agreed upon if they do not respect and trust one another. Identity conflict leads to an "us" versus "them" stance; only by humanizing people to one another can that identity conflict be reconciled. Yet, the importance of quality relationships among diverse individuals and groups is often undervalued or overlooked in conservation. Positive, constructive relationships, especially between members of disparate groups, are critical to catalyzing positive social change (Wheatley, 1998; Lederach, 2005).

An effective process is responsive to the relationship-building needs for dignity, respect, and trust among stakeholders. A good process provides the opportunity for unmet social and psychological needs to be met and for individuals to reconcile fractured relationships. This in turn supports creativity and the development of well-supported solutions. Too often, however, decision-makers rush to make decisions through processes that fail to engage all levels of conflict and thus fail to provide an adequate space to foster positive, durable relationships. A conservation conflict transformation process, by contrast, begins well before a multistakeholder meeting is designed or a decision point is targeted. It starts with an inclusive engagement with diverse individuals to better understand the conflict dynamics within the system from diverse perspectives. There is a continual focus on relationships, thereby creating the conditions for widely supported and durable decision-making.

Conservation conflict transformation practitioners rely on a replicable set of principles, skills, theories, and processes to ensure that process, relationship, and the substance elements of a conflict, as well as all levels of conflict, are addressed. One of the best practices in transforming conflict and sustaining conflict transformation mechanisms involves building the capacity of conservation teams and diverse stakeholders to do just that (Madden and McQuinn, 2014). Capacity building builds awareness among stakeholders of their role in creating or perpetuating destructive conflict, as well as their power to transform it constructively. Embedding the capacity to transform conflict among the individuals and groups within the conflict system empowers these people and groups to continually anticipate and address conflict in productive ways. The capacity-building process is also a secure and neutral opportunity in which previously antagonistic actors can become humanized to one another, create and build on "small wins," foster trust, and become motivated to work constructively together for positive change (Brown, 2003; Ansell and Gash, 2008).

1.4 Conclusion

Conservation conflict transformation is an approach that, in addressing all levels and sources of conflict, matches the reality of today's conservation conflicts. Conservation conflict transformation views conflict as an opportunity. As an approach, conservation conflict transformation focuses on reconciling destructive relationships through constructive change processes that address conflict at micro and macro scales. By ensuring that the deeper social and psychological needs, as well as the tangible, physical issues, are addressed, conservation conflict transformation offers a greater possibility for a good outcome. Conservation conflict systems are continually evolving, just as are the ecosystems in which they are embedded. The approach of conservation conflict transformation is equally dynamic, thus ensuring that conservation efforts can anticipate and adapt more effectively to the ongoing changes in social and ecological systems.

Successful incorporation of conservation conflict transformation begins with analysis. The focus of this chapter and the thematic inclusion of levels of conflict throughout this book hint at both the importance and the adaptability of conservation conflict transformation for a variety of marine-based conservation conflicts. As a first step in the analytic process, a levels-of-conflict analysis gives the practitioner a broader view and deeper sense of the severity of the conflict and provides insights into how strategies can be developed for the conflicts' effective transformation.

Capacity building in conservation conflict transformation, a focus on relationships, and a reconceptualization of what it means to successfully engage stakeholders in decision-making processes are critical changes needed within the marine conservation field to ensure that conservation's long-term goals are supported and realized. The Human–Wildlife Conflict Collaboration (HWCC) is leading efforts to integrate conservation conflict transformation in both marine and terrestrial wildlife conservation efforts. As many of the chapters in this book demonstrate, there is a growing list of individuals and organizations that recognize the value of conservation conflict transformation and the positive impacts it can have for people and marine life. As our community of practice expands and deepens, we anticipate more opportunities to share and demonstrate lessons learned and participate in the further evolution of this budding discipline.

References

Anderson, M. B., Olson, L., and Doughty, K. (2003). *Confronting War: Critical Lessons for Peace Practitioners.* Cambridge, MA: The Collaborative for Development Action, Inc.

Ansell, C. and Gash, A. (2008). Collaborative governance in theory and practice. *Journal of Public Administration Research and Theory*, **18**(4):543–71.

Atran, S. and Axelrod, R. (2008). Reframing sacred values. *Negotiation Journal*, **24**(3):221–46.

Balint, P., Stewart, R., Desai, A., and Walters, L. (2011). *Wicked Environmental Problems: Managing Uncertainty and Conflict.* Washington, DC: Island Press.

Barlow, A. C. D., Greenwood, C. J., Ahmad, I. U., and Smith, J. L. D. (2010). Use of an action-selection framework for human–carnivore conflict in the Bangladesh Sundarbans. *Conservation Biology*, **24**(5):1338–47.

Blackstock, K., Kelly, G., and Horsey, B. (2007). Developing and applying a framework to evaluate participatory research for sustainability. *Ecological Economics*, **60**(4):726–42.

Brown, M. M. (2003). Democratic governance: toward a framework for sustainable peace. *Global Governance*, **9**(2):141–6.

Burgess, H. (2004). *High-Stakes Distributional Issues.* <http://www.beyondintractability.org/essay/distribution-issues>, accessed March 4, 2015.

Burton, J. W. (1984). *Global Conflict.* Brighton: Wheatsheaf.

Burton, J. (1990). *Conflict: Basic Human Needs.* New York: St. Martin's Press.

Burton, J. (1993). *Conflict: Human Needs Theory.* New York: St. Martin's Press.

Canadian Institute for Conflict Resolution. (2000). *Becoming a Third-Party Neutral: Resource Guide.* Ottawa: Canadian Institute for Conflict Resolution.

Clark, M. (1990). Meaningful social bonding as a universal human need. In Burton, J. W. (ed.). *Conflict.* New York: St. Martins Press, pp. 34–59.

Clark, T. W. (2002). *The Policy Process: A Practical Guide for Natural Resource Professionals.* New Haven, CT: Yale University Press.

Coate, R. A. and Rosati, J. A. (1988). *The Power of Human Needs in World Society.* Boulder, CO: Lynne Rienner Publishers.

Coleman, P. T. (2011). *The Five Percent: Finding Solutions to Seemingly Impossible Conflicts.* New York: PublicAffairs.

Conover, M. R. (2001). *Resolving Human–Wildlife Conflicts: The Science of Wildlife Damage Management.* Boca Raton, FL: CRC Press.

DeCaro, D. and Stokes, M. (2008). Social-psychological principles of community-based conservation and conservancy motivation: attaining goals within an autonomy-supportive environment. *Conservation Biology*, **22**(6):1443–51.

Decker, D. J., Riley, S. J., and Siemer, W. F. (2012). *Human Dimensions of Wildlife Management*, 2nd edn. Baltimore, MD: The John Hopkins University Press.

Deutsch, M. (1973). *The Resolution of Conflict.* New Haven, CT: Yale University Press.

Deutsch, M. and Coleman, P. T. (2012). *Psychological Components of Sustainable Peace.* New York: Springer.

Deutsch, M., Coleman, P. T., and Marcus, E. C. (2006). *The Handbook of Conflict Resolution: Theory and Practice.* San Francisco, CA: Jossey-Bass.

Dickman, A. J. (2010). Complexities of conflict: the importance of considering social factors for effectively resolving human–wildlife conflict. *Animal Conservation*, **13**(5):458–66.

Doucey, M. (2011). Understanding the root causes of conflicts: why it matters for international crisis management. *International Affairs Review*, XX(2). <http://iar-gwu.org/node/393>, accessed March 5, 2015.

Dukes, E. (1999). Why conflict transformation matters: three cases. *Peace and Conflict Studies*, **6**(2):47–66.

Fisher, R., Ury, W., and Patton, B. (1991). *Getting to Yes: Negotiating Agreement Without Giving In*. Boston: Houghton Mifflin Harcourt.

Galtung, J. (1990). International development in human perspective. In Burton, J. W. (ed.). *Conflict*. New York: St. Martin's Press, pp. 301–35.

Ginges, J., Atran, S., Medlin, D., and Shikaki, K. (2007). Sacred bounds on rational resolution of violent political conflict. *Proceedings of the National Academy of Sciences of the United States of America*, **104**(18):7357–60.

Hendrick, D. (2009). *Complexity Theory and Conflict Transformation: An Exploration of Potential and Implications*. Centre for Conflict Resolution, Department of Peace Studies, Working Paper 17, June 2009, University of Bradford.

Jeong, H.-W. (2008). *Understanding Conflict and Conflict Analysis*. London: Sage Publications.

Lederach, J. P. (1997). *Building Peace: Sustainable Reconciliation in Divided Societies*. Washington, DC: United States Institute for Peace.

Lederach, J. P. (2003). *Little Book of Conflict Transformation*. Intercourse, PA: Good Books.

Lederach, J. P. (2005). *The Moral Imagination: The Art and Soul of Building Peace*. Oxford: Oxford University Press.

Lederach, J. P., Neufeldt, R., and Culbertson, H. (2007). *Reflective Peacebuilding: A Planning, Monitoring and Learning Toolkit*. Notre Dame, IN: The Joan B. Kroc Institute for International Peace Studies, University of Notre Dame and Catholic Relief Services.

Leong, K. M., Emmerson, D. P., and Byron, R. (2011). The new governance era: implications for collaborative conservation and adaptive management in Department of the Interior Agencies. *Human Dimensions of Wildlife*, **16**(4):236–43.

Leong, K. M., Forester, J. F., and Decker, D. J. (2009). Moving public participation beyond compliance: uncommon approaches to finding common ground. *The George Wright Forum*, **26**(3):23–39.

Levinger, M. (2013). *Conflict Analysis: Understanding Causes, Unlocking Solutions*. United States Institute of Peace Academy Guides. Washington, DC: USIP Press.

Lovallo, D. and Sibony, O. (2010). The case for behavioral strategy. *McKinsey Quarterly*, **2**:30–43.

Madden, F. (2004). Creating coexistence between humans and wildlife: global perspectives on local efforts to address human–wildlife conflict. *Human Dimensions of Wildlife*, **9**(4):247–57.

Madden, F. and McQuinn, B. (2014). Conservation's blind spot: the case for conflict transformation in wildlife conservation. *Biological Conservation*, **178**:97–106.

Manolis, J. C., Chan, K. M., Finkelstein, M. E., et al. (2009). Leadership: a new frontier in conservation science. *Conservation Biology*, **23**(4):879–86.

Marker, S. (2003). *Unmet Human Needs*. <http://www.beyondintractability.org/essay/human-needs>, accessed March 5, 2015.

Maslow, A. (1954). *Motivation and Personality*. Reading: Addison-Wesley Publishing Company.

Max-Neef, M. A., Elizalde, A., and Hopenhayn, M. (1989). *Human Scale Development: Conception, Application and Further Reflections*. New York: Apex.

Miall, H. (2004). *Conflict Transformation: A Multi-Dimensional Task*. <http://www.wageningenportals.nl/msp/resource/conflict-transformation-multi-dimensional-task>, accessed March 5, 2015.

Michalski, F., Boulhosa, R. L. P., Faria, A., and Peres, C. A. (2006). Human–wildlife conflicts in a fragmented Amazonian forest landscape: determinants of large felid depredation on livestock. *Animal Conservation*, **9**(2):179–88.

Pearce, W. B. and Littlejohn, S. W. (1997). *Moral Conflict: When Social Worlds Collide*. Thousand Oaks, CA: Sage Inc.

Peterson, M. N., Peterson, M. J., Peterson, T. R., and Leong, K. (2013). Why transforming biodiversity conservation conflict is essential and how to begin. *Pacific Conservation Biology*, **19**(2):94–103.

Redpath, S. M., Young, J., Evely, A., et al. (2013). Understanding and managing conservation conflicts. *Trends in Ecology and Evolution*, **28**(2):100–9.

Reed, M. (2008). Stakeholder participation for environmental management: a literature review. *Biological Conservation*, **141**(10):2417–31.

Rothman, J. (1997). *Resolving Identity-Based Conflict: In Nations, Organizations, and Communities*. San Francisco, CA: Jossey-Bass.

Sites, P. (1990). Needs as analogues of emotions. In Burton, J. W. (ed.), *Conflict: Human Needs Theory*. New York: St. Martin's Press, pp. 7–33.

Treves, A., Wallace, R. B., and White, S. (2009). Participatory planning of interventions to mitigate human–wildlife conflict. *Conservation Biology*, **23**(6):1577–87.

Wheatley, M. (1998). The promise and paradox of community. In Hesselbein, F., Goldsmith, M., Beckhard, R., and Schubert, R. F, *The Community of the Future*. New York: The Peter Drucker Foundation for Nonprofit Management, pp. 9–18.

Section 2

POLICY AND HUMAN–WILDLIFE CONFLICT

2

Conservation on Island Time

Stakeholder Participation and Conflict in Marine Resource Management

Catherine Booker and d'Shan Maycock

2.1 Introduction

The Bahamas has a relatively short but progressive history of enacting laws and policies to promote marine conservation. The earliest notable action was in 1958, with the establishment of the first national park in the country and the Caribbean region; the park later became the first fully protected marine reserve in the Caribbean region (Bahamas National Trust, 2013). In 1977, the Fisheries Resources Act was introduced, primarily to declare an exclusive fisheries zone during negotiations of the Third United Nations Conference on the Law of the Sea (1973–1982; UNCLOS III; Craton and Saunders, 1998) and to create the Department of Marine Resources (DMR). The DMR was mandated "to administer, manage, and develop the fisheries sector as stipulated by the Fisheries Resources," and to enforce fisheries regulations, marine mammal regulations, and seafood processing and inspection regulations (Government of The Bahamas, 2011). Efforts to address issues of marine pollution and environmental health began with the introduction of the Environmental Health Services Act (1987). In 1989, the Bahamas Environment Science and Technology Commission (BEST) was created "to protect, conserve, and manage the environmental resources of The Bahamas." BEST's mandates include coordination of activities related to international conventions and protocols, and responsibility for proposing environmental policy. The Bahamas participates in a variety of international conventions, such as the United Nations Convention on Biological Diversity, Wetlands, and Climate Change, which guide the country's environmental and species protection laws and policies and are relevant to marine conservation. More recently, together with Grenada, the Bahamas launched the Caribbean Challenge in 2008, leading the region in a commitment to protect 20% of its nearshore and coastal environments by 2020 (Broad and Sanchirico, 2008). Encouraged by the country's increasingly influential environmental NGO (eNGO) community, the government has also taken significant steps to protect commercially

Human–Wildlife Conflict: Complexity in the Marine Environment. Edited by Megan M. Draheim, Francine Madden, Julie-Beth McCarthy, and E. C. M. Parsons © Oxford University Press 2015. Published 2015 by Oxford University Press.

important marine species including sea turtles, sharks, Nassau grouper (*Epinephelus striatus*), and spiny lobster (*Panulirus argus*), as well as entire ecosystems within protected area boundaries (Bahamas Department of Marine Resources, 2010; Bahamas National Trust, 2011; Knapp et al., 2011).

The movement to protect and manage Bahamian marine environmental resources has progressed rapidly over the past 50 years; however, as might be expected, it has not always been an easy road for Bahamian conservationists, policy-makers, or resource users. A rapidly growing tourism industry and the development of a lucrative commercial fishing industry have raised the costs of conservation "success" and resulted in the perception that protection of the environment is in direct competition with the economic interests of communities throughout the country. Perhaps, as in other countries in the wider Caribbean, an inadequate understanding of socioeconomic issues and stakeholder perceptions has hindered the effort to gain community support for greater protection of the marine environment (Mascia, 1999; Broad and Sanchirico, 2008). Certainly, disagreement between resources users and conservationists, over the value, use, and management of fisheries and coastal areas, has followed the larger global pattern in biodiversity and natural resource conservation, making conflict over these issues one of the most difficult problems facing practitioners today (Young et al., 2010; Redpath et al., 2013). As the Bahamas continues its effort to balance conservation of its most valuable marine resources and the livelihoods of its citizens, more emphasis on the human dimensions of conservation to ensure sustainable outcomes will be of critical importance, especially considering the inherent potential for conflict when momentum for conservation interventions is accelerated.

In this chapter, we present two case studies, from the most valuable marine fishery in the Bahamas and from a unique segment of the tourism industry, to explore the emergence of conflict over proposed and existing laws during multistakeholder participatory processes. We analyze the conflicts in each case through the lens of the levels of conflict model introduced by HWCC (Chapter 1; HWCC, 2008; Madden and McQuinn, 2014) and then use best participatory practices identified by Reed (2008) and others (Chess and Purcell, 1999; Stringer et al., 2006; Luyet et al., 2012) to compare the stakeholder engagement approaches used in each scenario. While we present no quantifiable data to demonstrate the relationship between the use of good participatory practices and the degree of conflict during and after the conservation interventions, we believe that the contrast in the practices used in both scenarios provides a good opportunity to discuss those practices in the context of conflict transformation. Conservation conflict transformation is an approach that acknowledges the social and psychological aspects of conservation conflict and "aims to achieve durable positive coexistence, not simply short term or superficial compliance" in conservation interventions (Madden and McQuinn, 2014). The cases are presented from the perspective of the practitioners who played leadership roles in those conservation efforts and thus may provide "lessons learned" for those involved in fisheries and water quality conservation initiatives facing similar challenges.

2.2 Case study: tourism and community-based harbor management in Exuma

2.2.1 Background

Like most Small Island Developing States (SIDS), the tourism industry in the Bahamas is of critical importance to both national and local economies. Yachting tourism, a small but significant segment of the tourism market, is especially important on the less accessible and therefore less visited Family Islands in the archipelago. Geographically, the Family Islands are spread throughout the island chain and are ideal destinations for yachting tourists, also known as cruisers, seeking an "off the beaten path" experience. The growing number of cruising vessels has indeed had a beneficial economic impact in the most visited Family Island communities but also presents new environmental management challenges (Sullivan-Sealey, 2000).

The Exuma Cays are located in the south central Bahamian archipelago and are considered to be some of the most beautiful cruising grounds in the world. Elizabeth Harbour, located adjacent to the Family Island settlement of George Town, Great Exuma (Figure 2.1), is the most popular cruising destination in the Exuma Cays. This large, natural harbor is visited by hundreds of cruising tourists every year during the winter and spring months that make up

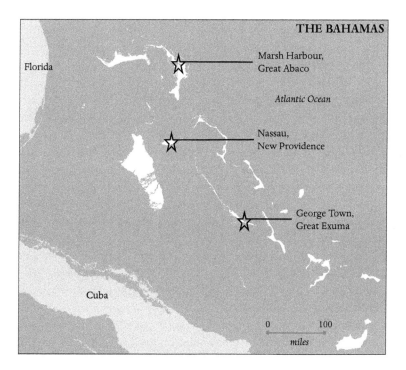

Figure 2.1 The Bahamian archipelago.

the height of the tourism season. Some of the tourists spend a few days in the harbor as they pass through en route to the Caribbean but most stay for months at a time. Cruising tourists are drawn to Elizabeth Harbour for its conveniences, natural beauty, and weather protection, as well as opportunities to socialize with other cruisers (Sullivan-Sealey, 2000; Booker, unpublished observations). A perceived freedom from regulation and the ability to maintain a low cost of living also make the area attractive to a segment of this tourism market.

Great Exuma is home to approximately 7,000 people, with George Town as its largest settlement (Bahamas Department of Statistics, 2012). The majority of the island's population is Bahamian, but there is also a seasonal expat community with homes near George Town along the coastline of Elizabeth Harbour. Both communities enjoy the harbor's numerous beaches, but seasonal residents have better access to its waters. Elizabeth Harbour is the focal point of tourism and development activity and is recognized as an invaluable natural asset to the island's economy.

In the mid-1980s a small group of local citizens and expats residing on Great Exuma became concerned about the degradation of reefs and fish populations in Elizabeth Harbour and attributed this change to overfishing and sewage pollution. Direct discharge of wastewater from cruising vessels was a growing concern, as the floating community in the harbor seemed to increase in number each year, at its peak reaching approximately 500 vessels. A formal report was made to the government in the mid-1990s by a marine biologist and dive-shop owner. In the late 1990s, a scientific water quality study was completed and published in *The Bahamas Journal of Science* (Sullivan-Sealey, 2000). This study, along with a harbor resource plan commissioned by the Ministry of Tourism in 2000 (MacGregor, 2000) and the voiced concerns of community leaders, allowed the Bahamian government and partners to win a Full Size Project grant from the Global Environment Facility (GEF) to address the issue of yacht-based pollution. Elizabeth Harbour, Exuma, became one of 14 demonstration projects throughout the Caribbean in the Integrating Watershed and Coastal Area Management (IWCAM) project, an effort designed to assist and encourage SIDS to adopt integrated watershed and coastal area management programs. Because of its large yachting community, Exuma qualified as an ideal location to demonstrate how the installation of pumpout facilities to remove sewage waste from boats (Figure 2.2), waste treatment to properly dispose of that waste, and moorings to protect the seabed from anchoring, could improve water quality in harbors throughout the Bahamas and the Caribbean and bring social, economic, and environmental benefits to local communities. The longer-term objectives of the project included changes in environmental law and policy to support new wastewater management operations, and a sustainability plan to ensure continued financial and local support of the initiative (IWCAM, 2013). Documents prepared by the environmental consultant (Booker, unpublished) and final case studies for the project (Clauzel, 2011; UNEP-CAR/RCU, 2014) are used in this document to describe the scenario.

At the beginning of the five-year implementation of the IWCAM project, the BEST Commission formed the Elizabeth Harbour Management Steering Committee (EHMSC), with the intention of encouraging community participation in the decision-making process. Members of the EHMSC represented local and central governments, as well as Exuma business and community members; however, a primary stakeholder group, the cruising community, was

Figure 2.2 The pumpout boat visits a yacht in Elizabeth Harbour.

not represented on the committee. Early on, the Exuma IWCAM project was faced with delays, and efforts to communicate the objectives of the project to stakeholder groups were suspended. As a result, the EHMSC was inactive for the majority of the first years of the project. The project was re-energized when BEST contracted a project consultant (C. Booker) to provide full-time support in Exuma for the EHMSC. Project activities resumed, but with a greatly reduced time frame, to meet objectives and a new challenge: a growing resistance within the cruising community to the proposed introduction of pumpout services and moorings.

The EHMSC and the IWCAM project found early success in their additional efforts to gain local community support and participation. An enterprising Exumian was granted permission by the local authorities and the EHMSC to operate a pumpout service to collect sewage waste from the boats and was set to begin before the end of the cruising season. After the EHMSC gained a grasp of the tasks at hand, a meeting was held with several cruising community leaders to discuss the primary goals of the IWCAM project. Unfortunately, the cruisers felt targeted by the aims of the project to change their current practice of direct-sewage discharge into the harbor, and a complex conflict scenario began to unfold, placing the two primary stakeholder groups, the cruisers and the EHMSC, against each other (discussed later in the case conflict analysis). Few boats used the pumpout service when it was introduced later in the season, and the majority of the cruisers did not take the no-discharge law seriously. Without recognizing the longer-term consequences of ignoring the brewing conflict and faced with constant logistical challenges and a looming project deadline, the EHMSC moved forward with implementation of the project infrastructure objectives during the summer off-season in the absence of the cruising community. During this time, moorings were installed and the construction of a wastewater treatment facility was completed. The EHMSC also drafted a harbor management plan, made policy recommendations to strengthen and adapt existing environmental laws, and created sustainability plans for new harbor infrastructure and operations. An important decision to form a nonprofit company that would formalize the community's commitment to cleaning up the harbor and provide a platform for better stake-holder participation in the future was stated in these plans, but this entity would not come into being until almost two years after the conclusion of the IWCAM project.

The cruising community returned to Elizabeth Harbour at the start of the following winter season to find that the IWCAM project had moved forward despite their lack of support. Once again, the tensions from the previous season rose, building on the unresolved issues between the two communities. This time, the EHMSC was challenged not only with the issue of low compliance, but even with active efforts to undermine the project by a small but vocal and influential group of the cruising community, who encouraged others to boycott the moorings and pumpout services and question local authorities about the validity of the application of existing environmental laws to support a "no-discharge" regulation in the harbor. Eventually, supporters of the project emerged from the cruising community and made clear that those dissenting voices were not representative of their views of the effort; but, as the complexity of the situation increased, the EHMSC was left on shaky ground. As the EHMSC was a new and not yet formalized entity, with limited resources and authority, it was perceived by commit-tee members that they had little choice but to defend their decisions and hope that eventually the cruising community would accept their new harbor management policies.

2.2.2 Conflict analysis

The formation of a cruising community in Elizabeth Harbour began in the early 1980s, when George Town became a popular cruising destination. A mutually tolerant relationship between the two communities initially existed, with Exumians benefiting from an influx of tourist dollars and charitable contributions, and the cruisers benefiting from conveniences of a friendly local population that showed little concern for their activities in the harbor. The first signs of discord between the cruising community and the local community occurred in the late 1990s, when signs of environmental degradation in the Elizabeth Harbour started to become more apparent. By the late 2000s, a relatively benign dispute over who was at fault for the environmental degradation of Elizabeth Harbour had progressed into an underlying and identity-based conflict that tested the friendly relationship between the communities.

2.2.2.1 *The dispute*

The original dispute between the local community and the cruising community formed over two decades but came to the forefront of public debate when the aims of the IWCAM became publicly known. The new dispute was focused on the goals of the project. Did IWCAM goals, the imple-mentation of moorings and pumpout services, appropriately address the environmental issues in Elizabeth Harbour and take into account the multiple reasons for declining water quality in the harbor? Or was the burden of blame placed unfairly on one group, the cruising community?

2.2.2.2 *Underlying conflicts*

The dispute that emerged during the IWCAM project was rooted in two primary underlying conflicts between the cruising community and the EHMSC. The first and most obvious underlying conflict centered on the historical dispute over the cause of the pollution and ecosystem degradation in the harbor. If a cruiser were asked, the coral reefs in the harbor were dying because of leaky septic systems and coastal development. A resident of George Town would most likely say it was because of the many thousands of gallons of boat sewage that directly discharged into harbor waters by the cruising yachts. In reality, both answers were probably correct to some extent; but the urge to not take responsibility and blame the other party was acted on by the most vocal members of both groups once the project began. Indeed, it seemed that the aim of the project was clearly directed at the cruising community, as it was they who would be required to change their behaviors and comply with new harbor regulations.

Because of the historical underlying conflict over responsibility for pollution in the harbor, it was not surprising that the Exuma community quickly accepted the goals of the IWCAM project, while long-time members of the cruising community felt as if they were being unjustly blamed for the problem. Given the little information with which they had to work because of the delays in the project communications program, the loudest critics of the EHMSC began to question the validity and motivation for the Exuma IWCAM project. Emotions ran high during a town-hall meeting arranged by the EHMSC, as cruising community members stood up to challenge the science behind the perceived water quality issue and presented their own "evidence" that the impact on the water quality was negligible compared to what was coming from the toilets and laundry facilities on Great Exuma.

The second underlying conflict was less appreciated by the EHMSC because it originated from previous experiences that veteran cruisers in the harbor had had with laws and regulations in their home countries (see Comment). The experiences of some members of the cruising community with no-discharge and no-anchoring laws in the United States and Canada seemed to also be at the root of a preexisting negative perception of such regulations in Elizabeth Harbour. Based on these previous experiences, assumptions were made about how expensive, unsafe, and unsanitary the moorings and pumpout services would be if brought to Elizabeth Harbour. Even though the frustration felt by the boaters originated in another geographic location and because of another entity's actions, the existing resentment translated into underlying conflict with the EHMSC and significantly affected the IWCAM project.

Comment

For another example of concern over top-down regulations, see Sarah Wise's discussion of MPA creation in the Bahamas (Chapter 6). Interestingly, in Wise's example, the concerned parties were local residents of Andros Island, while in this chapter, those most concerned with such regulations are not locals but rather temporary residents and tourists.

—The Editors

2.2.2.3 Identity-level conflicts

Two identity-based conflicts were not vocalized as openly during the public debate surrounding the IWCAM project but were still influential in some of the more important decisions made by the EHMSC and actions taken by the cruising community.

Because there was very little forum for communication between the two groups, members of the cruising community and the EHMSC would often explain the other group's behavior and stance on the IWCAM project using stereotypes and assumptions. The lack of rapport between the groups made it was easy to make assumptions and to use stereotypes to characterize the other's identity and motivations. For example, it was assumed by some members of the EHMSC that the cruisers were not interested in supporting the project because they were "cheap" and "it was not in their budget." The cruisers expressed concern that Exumians did not have the capacity either in training or experience to install and maintain the project infrastructure and that therefore the moorings, for example, could not be trusted. These assumptions undermined trust and kept positive and constructive dialog about some legitimate concerns from taking place. It was also the basis of identity-based conflict between the groups.

In a broader context, the perception that the IWCAM project was a top-down, government-driven project was in direct conflict with the identity of a segment of the cruising community who were living on their boats and cruising in the Bahamas to escape government regulation. To the frustration of the EHMSC, rumors circulated that fed these unfounded fears, including a U.S. government conspiracy.

2.3 Case study: fishery sustainability and behavior change in Abaco

2.3.1 Background

Like the tourism industry, the fishing industry is of critical importance to the economies of Caribbean SIDS such as the Bahamas. Fishing traditionally provides employment for less skilled citizens and can be an extremely profitable endeavor for more organized commercial operators. The spiny lobster (*Panulirus argus*) fishery has been a cornerstone for the economic development of the Bahamas since the early 1950s, when live lobsters were exported to Florida by the Agriculture and Marine Products Board (Craton and Saunders, 1998). By the early 1970s, the industry began marketing lobsters as frozen tails, which is still the preferred export method today. In 2006 and 2008, the total export value of lobster tails was $77 million and $92 million, respectively, making the fishery by far the most lucrative in the country (Bahamas Department of Marine Resources, 2010). The high export value of lobster tails continues to be fueled by demand from the countries of Europe, Canada, and the United States.

In the Family Islands of Abaco, located in the northwest sector of the Bahamas (Figure 2.1), the spiny lobster fishery is considered a key driver of the local economy alongside the growing tourism industry. The population of "the Abacos" is approximately 15,000 people, with most living on the islands of Great Abaco and Little Abaco (Bahamas Department of Statistics, 2012). The main settlement of Marsh Harbour is a hub of tourism and development activity, but many communities outside of Marsh Harbour depend on fishing as a primary source of income. Abaco is considered one of the top three islands in the Bahamas for lobster landings and exports.

In 2009, a local NGO, Friends of the Environment (FRIENDS), partnered with an international conservation organization, Rare,[1] to engage Abaco communities in a multistakeholder process to identify the organization's next conservation project. The implementation of the campaign was captured in Rare project documents that are used to describe this case (Maycock, 2010).

Before building the basis for a Rare signature "Pride" campaign, FRIENDS's education coordinator (D. Maycock) was trained in social marketing methodology and facilitation. She then organized several community workshops to determine what Abaconians considered to be the top threats facing the marine environment. Fishermen, resource users, marine resource managers, local government officials, teachers, environmental groups, seafood wholesalers, and restaurant owners were invited to attend the workshops in an effort to include a variety of perspectives on the issues and represent the diverse communities of Abaco (Figure 2.3). Because of the importance of the lobster fishery to the local economy, it was not surprising when stakeholders identified unsustainable lobster fishing practices as

Figure 2.3 Size Matters campaign meeting in Abaco.

---

[1] Rare is an international conservation organization focused on empowering local communities to solve conservation issues. Its signature method, the Pride Campaign, uses social marketing approaches and depends on partnership and training of a local conservation group; see <http://www.rare.org>.

the most critical threat to a marine species and therefore to the marine environment. Despite the existence of lobster fishery laws and the international pressure on the industry to maintain a higher quality product, a high incidence of illegal harvest of undersized lobsters had been a concern in Abaco for a number of years.

In response to the community's consensus that unsustainable fishing was a major threat to the marine environment and particularly their most lucrative fishery, FRIENDS began to work with key stakeholders, particularly lobster fishers, to develop the Pride campaign. The initial development phase of the campaign began with a process to identify the causes of this particular threat and the barriers to the removal of the threat. Focus group meetings and hundreds of personal interviews were conducted to further understand the issue. The data from meetings and interviews were then used to identify fishers as the primary target audience and begin to form campaign objectives. During this participatory process it became clear that stakeholders—including fishers—believed current fishery laws that set a minimum size for lobster tails and required the fishery be closed for four months of the year were adequate management measures, and that new laws were not necessary to solve the problem. It was thus decided that focusing on encouraging voluntary compliance to existing laws by changing the behavior of the fishers through a series of social marketing activities would be the most practical approach. The effort became known as the "Size Matters" campaign, to emphasize the message that fishing undersized lobsters was no longer acceptable.

Initially, input into the Size Matters campaign came mainly from the primary stakeholder group, the fishers; but it wasn't long before another group, the local buyers and exporters, became eager to get involved. The participatory approach allowed for their inclusion in the process; but subsequent consultations revealed that while the exporters were glad the campaign was underway, they were also doubtful of the impact it could have. They too had a history of conflict with the fishers and felt that their own efforts to educate them had not been successful.

During a series of stakeholder meetings, the exporters openly voiced their concern about the quality of their product and the impact fishers' practices were having on their ability to sell to outside markets. Local buyers and exporters agreed that the high incidence of undersize lobster fishing was a major concern and one that was difficult to manage; but in addition, they revealed their worry that fishers were also using unsafe practices in their handling of lobster catches. In fact, the exporters argued that fishers' disregard of size limits and unsafe handling practices were a direct threat to a national effort underway to achieve an eco-label certification for the fishery, and therefore a direct threat to their industry. If they were not able to comply with the new product standards, it could have a devastating impact on their access to the European and American markets. This shared bit of information triggered recognition of the real issue and a turning point in the campaign, when in the middle of a heated meeting a fisherman stood up and said, "At the end of the day, the only regulation we have is our conscience!" Finally, the fishers were able to understand the exporters' perspective, and the direct impact of their behavior on the fishery as a whole.

Recognizing an opportunity to encourage further cooperation between the two groups, the education coordinator organized facilitated meetings in the summer of 2010 to bring the fishers and exporters together again to find a shared solution to the problem. The influence of the exporters was clear, as fishers depended on them to buy 95% of their catches; but the realization of the fishers' responsibility to adopt the solution was an important step in reaching an agreement. The exporters offered to compromise and provide the incentives asked for by the fishers, such as providing ice, and measuring tools or gauges. In some cases, the exporters agreed to pay a higher price to those that remained in compliance.

After lobster season opened in 2010, top exporters reported that their receipts of undersize catches from fishermen were almost zero. They attributed this desired result to the participatory process and message of the Size Matters campaign. The lobster-tail gauge developed during the campaign is now used by fishers throughout the Bahamas and is an example of a successful effort to expand the use of a feasible, low-enforcement management strategy (Figure 2.4).

2.3.2 Conflict analysis

While the participatory approach pursued in developing the Pride campaign was essential in identifying the appropriate target audience and message the campaign would pursue, it also helped to reveal the different levels of conflict contributing to the issue of waning compliance in the fishery. Like the Exuma case study, a history of unresolved disputes existed before any decisive conservation action was taken. However, even though underlying and identity-level conflicts did surface throughout the project, following the greater focus on the participation

Figure 2.4 Lobster gauge developed during the Size Matters campaign.

of all stakeholders during the development and implementation of the campaign allowed the education coordinator to more effectively manage conflict.

2.3.2.1 The dispute

The primary dispute in this case was not centered over the fact that the fishers were or were not compliant to fishery laws, but instead over the cause of fishers' non-compliance. This dispute surfaced early in stakeholder group discussions. Although the fishers were willing to recognize that too many undersized lobsters were being harvested, they were reluctant to take all of the responsibility for their unsustainable fishing practices. They claimed that they did not intentionally fish illegal sizes, but that this practice occurred because they lacked the proper tools needed to measure their lobster catches. They also explained that beneath the water, everything appeared larger, so it was easy to "mistakenly" fish the wrong sizes. Fishers pointed to the other key stakeholder groups, the exporters (those responsible for buying and selling the fishers' catch) and the DMR as being responsible for low compliance in the fishery.

2.3.2.2 Underlying level conflicts

The basis for the dispute over responsibility for the increasing harvest of undersized lobster was rooted in a history of insufficient enforcement of fishery laws, and the strained relationship between the fishers and the few buyers who controlled the prices paid for their catch.

When the Size Matters campaign formalized conversations between the fishers, buyers, and the DMR, these underlying level conflicts emerged as having a significant effect on the behavior of the fishers. For example, fishers held a long-time belief that the DMR was not likely to be an enforcement presence, which some felt effectively excused them from obeying the law. This belief was acted out when fishers accused the DMR for not executing their "job" of managing the fishery properly. In one meeting in Marsh Harbour, tensions rose as fishers voiced their concerns loudly to the DMR, and a back-and-forth argument between the two ensued. The mostly justified assumptions that the DMR lacked funding and capacity and consequently were unable to enforce many fisheries regulations was clearly a sticking point in this debate, making it unlikely that constructive dialog would be possible between the two groups.

Underlying conflict came into play in the fishers' relationship with the buyers because of the historical dispute over prices. The fishers were not likely to trust the exporters' motivation for seeking better compliance because they were already disappointed that they were not receiving sufficient compensation for their catches and believed that their ability to sell good-quality catches was hindered by the exporters not providing the necessary tools such as ice and a tool to measure lobster tails.

2.3.2.3 Identity-level conflict

One source of identity-level conflict in this case is based around the assumptions each group held that the other would behave in a certain way. Again, these assumptions were formed by a history of interactions that tended toward mistrust and noncooperation. The history of unresolved underlying conflict based on each group's negative perception of the others' behavior and trustworthiness had repeatedly undermined the trust that was necessary to come to any agreement in the past, and thus a cycle of mistrust existed. When the exporters and buyers became more involved in the participatory process during the Pride campaign, they did so uncertain that the effort would be effective. Because of their history of failed attempts to encourage better compliance, they assumed that the fishers' behavior would not change.

A second source of identity conflict came from the fishers' basic need for respect. Like fishermen throughout the world, the personal identities of the lobster fishers of Abaco are strongly shaped by their occupation (Thompson et al., 1983). It was therefore inevitable that they would be threatened by any attempt to limit their access to the fishery, particularly if this attempt was perceived as not taking into account their knowledge of the resource and their inherent reliance on this way of life (Couthard et al., 2011).

2.4 Discussion: a comparison of participatory processes and conflict transformation

A comparison of the participatory processes used to engage stakeholders in the harbor management case and in the lobster fishery case provides an opportunity to examine the practices and process factors that facilitated or hindered the transformation of underlying and identity-level conflicts and ultimately affected the outcomes of the conservation intervention. We rely on several literature reviews of participatory "best practices" by Reed (2008; Figure 2.5) and others (Chess and Purcell, 1999; Luyet et al., 2005; Stringer et al., 2006) to

- Stakeholder participation needs to be underpinned by a philosophy that emphasizes empowerment, equity, trust, and learning.
- Where relevant, stakeholder participation should be considered as early as possible and throughout the process.
- Relevant stakeholders need to be analyzed and represented systematically.
- Clear objectives for the participatory process need to be agreed among stakeholders at the outset.
- Methods should be selected and tailored to the decision-making context, considering the objectives, type of participants, and appropriate level of engagement.
- Highly skilled facilitation is essential.
- Local and scientific knowledge should be integrated.
- Participation needs to be institutionalized.

Figure 2.5 Principles of best-practice stakeholder participation (Reed, 2008).

frame the discussion and compare the practices observed in the case studies. We focus our discussion on those practices that parallel a conservation conflict transformation approach in addressing the substance, process, and relationship sources of conflict (Madden and McQuinn, 2014). While we do not claim to identify a direct cause-and-effect relationship between the practices demonstrated in the cases and their effectiveness in creating or transforming conflict, we believe the contrast highlights the importance of a more thoughtful approach to participatory process planning, especially when voluntary compliance is necessary for conservation success.

2.4.1 Building relationships and trust

Arguably, the most important "best" participatory practices are those that help to build trust and relationships. Reed (2008) takes into account the need for a relationship-building process in his review of best participatory practices, which he describes as a "service contract" philosophy of stakeholder participation that emphasizes empowerment, equity, trust, and learning over a longer period of time. Processes that allow "social interactions among stakeholders, individual and group reflection on what is being learned, as well as iterative attempts to apply what is being learned to the issues under discussion" (Stringer et al., 2006), also known as "social learning" (Stringer et al., 2006; Evely et al., 2011), are more likely to result in the development of relationships, appreciation of others' trustworthiness or character, and adversarial relationship transformation realized by the stakeholders involved (Schusler and Decker, 2003). In scenarios where there is already a conflict situation or the potential for conflict is significant, this type of social learning is critical (Keen and Mahanty, 2006). While traditional participatory approaches may address and attempt to resolve relatively uncomplicated disputes and underlying conflicts, a conflict transformation approach delves into the more difficult identity-level conflicts by seeking to understand and transform relationships (Lederach et al., 2007; Madden and McQuinn, 2014). The conservation conflict transformation approach emphasizes "equal attention to the dialogue and relationship building needed to foster dignity, respect and trust, as well as support for more effective decision-making around tangible solutions" (Madden and McQuinn, 2014).

In the context of the case studies presented in this chapter, the success of the participatory processes differed noticeably in building relationships and fostering trust among the key stakeholder groups. The aim of the Exuma IWCAM project to include local stakeholders through the creation of a committee was a good step to promote a sense of empowerment and inclusion in the decision-making process within the local community; however, by excluding the cruising community from the committee, an opportunity to encourage social learning and inclusive decision-making that would have helped to build relationships and trust among members of the two groups was missed. Indeed, the time that was saved in the short term by not including cruising community representatives in the decision-making process was likely lost in the longer term when marginalized members of the cruising community attempted to obstruct the project by boycotting the moorings and pumpout service.

In contrast, the Size Matters participatory process empowered the key stakeholder group, the fishers, to not only help to identify the primary threat facing their industry but also to decide upon and adopt feasible and realistic solutions through negotiation with the exporters. Instead of blaming others, the fishers came to believe that their actions were the solution. Although meetings with fishers and exporters became heated at times, making the time and allowing flexibility during group meetings increased social learning through interpersonal communication. Opportunities were created for each group to understand the others' perception of the issue and to build trust among the groups. The result of the improved relationships between the fishers and exporters was a shared solution that worked for everyone.

2.4.2 Timing of stakeholder engagement

Engaging stakeholders early in the decision-making process, even before project implementation, is essential if participatory processes are to lead to high-quality and durable decisions (Reed, 2008; Luyet et al., 2012). It is more common that stakeholders are asked to take part in a process or project during or after project implementation in the format of the "big meeting"; however, this approach increases the risk that predetermined project goals may not represent stakeholder priorities or may even be at odds with the undiscovered needs of key groups (Madden and McQuinn, 2014). Motivating participants to contribute to decision-making may be difficult if they feel that their input will not be taken into consideration and that decisions have already been made (Chess and Purcell, 1999; Reed, 2008). Allowing flexibility for group reflection, learning, and adaptation of plans early and throughout the process may help to avoid these pitfalls (Chess and Purcell, 1999; Stringer et al., 2006).

The Size Matters campaign included a diverse stakeholder group from the onset and encouraged stakeholder participation in decision-making. Stakeholders contributed during the development and implementation phases of the campaign, and decisions were made with their input at each step. The adaptation of the Size Matters campaign to include the exporters demonstrated the flexibility of the participatory process.

In contrast, stakeholder participation in the Exuma IWCAM project was initiated as a result of local concerns but did not follow an inclusive approach early on in the decision-making process. The cruising community was not engaged in the process until after project implementation had begun and major decisions had already been made. As opposed to feeling included, the cruisers were placed in a reactive position that hindered cooperation and ultimately decreased compliance to new harbor regulations.

2.4.3 Stakeholder analysis and representation

It is widely accepted by conservation practitioners that stakeholder analysis should be a first step in the development of a participatory process. According to Reed (2008) "stakeholder analysis is a process that: (i) defines aspects of a social and natural system affected by a decision or action, (ii) identifies individuals or groups who are affected or can affect those parts of the system, and: (iii) prioritizes these individuals and groups for involvement in the decision-making process." A conflict transformation approach takes more traditional stakeholder analyses to another level by seeking "a deeper understanding of conflict dynamics at the interpersonal, community and systems levels" (Madden and McQuinn, 2014). An analysis of the levels of conflict, as explored in this chapter, and the conflict-intervention triangle framework adapted to anticipate the sources of conflict may be useful tools for conservation practitioners facing complex conflict scenarios (Madden and McQuinn, 2014). Comprehensive stakeholder and conflict analysis will lead to more productive participatory processes because stakeholders will be represented fairly. Stringer et al. (2006) state, "The diversity of stakeholders involved inevitably influences the success of a participatory project, as does their representativeness in terms of the issue or system concerned." Similarly, Luyet et al. (2012) conclude that while various approaches can be taken to identify stakeholders, it is critical that all stakeholders are included in a successful participatory process.

A formal stakeholder analysis tool was used to form the Size Matters campaign. Using input from multiple stakeholder groups, a matrix was developed that identified each stakeholder group for who they were, what they could offer, their ability to assist, and the risk of their nonparticipation. Inclusion of a diverse group of stakeholders in the initial stakeholder analysis process proved to be an important step in identifying the fishers as the target audience, and the exporters as a key influence group of the campaign. Ultimately, the inclusion of the fishers in the campaign led to more durable solutions to the problem. The development of a lobster-tail gauge was a direct result of the fisher's input, and a key reason why compliance increased.

A complete stakeholder and conflict analysis of both the cruising community and the Exuma community during the design of the Exuma IWCAM project would have helped project leaders to discover the key issues, concerns, and expectations of the groups, as well as the motivation each had to support or not support the project. A better understanding of group heterogeneity might have allowed the EHMSC to better gauge the "weight" of the adversarial stance taken by the cruising community leaders and to possibly identify supporters of the project earlier in the process. Potential conflicts might have been avoided if the underlying assumptions and trust issues, including the sense of marginalization felt by members of the cruising community, were identified and planned for earlier in the participatory strategy.

Appropriate representation of stakeholder groups in a participatory process is identified as a key factor in the outcome of a project (Reed, 2008), and the dilemma regarding the application of this practice in the Exuma IWCAM project was particularly interesting. Empowerment of local stakeholders is a recognized normative benefit of participation, as it supports democratic ideals and promotes active citizenship (Martin and Sherington, 1997); but at what level and to what extent was it appropriate to include a group without legal or democratic rights as a resource user in the decision-making process? The cruising community by definition was a primary stakeholder group, as the decisions made during the Exuma IWCAM project affected them most directly. They also possessed significant informal power (Madden, 2011) in that local enforcement limitations meant that their willingness to voluntarily comply with new harbor laws and policies was critical to the success of the project. If they continued to dis-charge waste and refused to use moorings in designated areas, environmental project object-ives would not be met and the financial sustainability of the project would be threatened. In addition, their impact on the tourism-based local economy afforded them influence on the acceptance of harbor management efforts among Exumians who depended on them as a customer base.

2.4.4 Skilled facilitation

Whether through the lens of public participation or conflict transformation, skilled planning and guidance in a conservation intervention is identified as vitally important to the effect-iveness of a participatory process and increases the likelihood of finding a workable solu-tion. The structure and methods chosen to engage stakeholders are also important; but it is the manner in which these methods are carried out that matters most, particularly in con-servation efforts when there is a high likelihood of conflict between resource managers and users (Chess and Purcell, 1999; Bojorquez-Tapia et al., 2004; Reed, 2008). As discussed in the previous sections, a conservation practitioner must have the experience and training to design participatory processes that function to build relationships, foster trust, and produce workable solutions for a diverse group of stakeholders. Madden and McQuinn (2014) describe the need for these kinds of "soft skills" in conservation, particularly when dealing with conflict in decision-making. They stress that such skills and strategies are no less important than expertise in the natural sciences when designing and implementing conser-vation interventions and also need to be learned and mastered. Experienced conservation practitioners must have the ability to remain impartial and open to different perspectives, while encouraging positive and equitable group dynamics, as well as recognizing and addressing conflict appropriately.

In both case studies, some level of facilitation skill was necessary to achieve the desired project outcomes. During the Exuma IWCAM project, the consultant led the EHMSC in the development of project deliverables including a harbor management plan and sustainability plans but did not have the resources or capacity to implement a more comprehensive partici-patory process which would have included the cruising community. While efforts were made to gather diverse stakeholder views, a more intentional and strategic approach might have achieved better results. As much as the consultant was able to encourage group participation and cooperation by the EHMSC members, it proved nearly impossible to manage the conflict with the cruising community not represented in this entity. From the perspective of conflict transformation, if disparities in decision-making power and the perceived lack of stakeholder influence are not addressed, as occurs frequently in poorly structured conservation interven-tions, conflicts may be exacerbated (Madden and McQuinn, 2014). In this case, the problems posed by the incomplete process presented additional challenges that were difficult for the consultant to overcome.

In contrast, the skill and strategic planning that was employed from the outset of the Size Matters campaign led to a more thoughtfully structured process. In the project identification phase, the education coordinator was empowered to design a process that would ensure project objectives were aligned and representative of the multiple stakeholder groups involved. The creation of a "safe" environment by a neutral third party allowed stakeholders who were mired in underlying and identity-level conflicts to openly discuss the issues that were having significant impact on their ability to problem solve. An awareness of timing during the process allowed the coordinator to find opportunities to make progress. Overall, the focus of the Size Matters campaign on community pride encouraged a sense of shared responsibility for the problem and for finding solutions in a cooperative manner.

2.5 Conclusion

Based on evidence from the policy, environmental conservation, and peace-building fields (Madden and McQuinn, 2014) and the case studies presented in this chapter, integration of meaningful stakeholder engagement in marine conservation is not just important but essential. The stakeholder participatory approach taken in the Size Matters campaign was effective in achieving the goal of greater fishery law compliance because fishers chose to change their behavior. This behavior change occurred because this key stakeholder group was included early on and empowered in a decision-making process. The process succeeded in bringing all stakeholders together in a constructive forum to allow transformation of an existing conflict that had made finding a common solution difficult. Clearly, prioritizing the inclusivity of the participatory process took more time and skill, but the outcome was well worth it.

In comparison, the unresolved underlying and identity-level conflicts between the cruising community and the EHMSC at the conclusion of the IWCAM project were a considerable obstacle to the project's success. The limited scope of the project set the stage for conflict by focusing on the impacts of one group while overlooking the potential negative impacts of other land-based resource users on water quality. The participatory process represented by the creation of the EHMSC was certainly inclusive of local interests and was successful in the longer term with the creation of a harbor-conservation nonprofit entity; however, because the process marginalized the stakeholder group that was most affected by the project, deeper distrust and support for improving water quality during project implementation was created, and opportunities for finding shared solutions may have been missed.

The Bahamas has made significant progress since the 1950s in protecting its marine resources. In order to continue to build on its legacy of conservation achievements and to ensure the sustainability of its efforts, increased consideration must be given to the inclusion of all resource users in strategic and inclusive participatory processes. Amid the pressure to meet international, regional, national, and local demands for greater protection of the marine environment, and the increasing pressure to develop coastlines and the tourism industry, policy-makers would be wise to increase budgets and time frames to improve the quality of the stakeholder participatory process rather than rush to meet objectives that create conflict and prolong the realization of important conservation goals.

Lessons learned

- Process is important. It is essential for conservation practitioners to obtain so-called soft skills such as participatory process design and that they recognize that these skills need to be practiced.
- Some parties in a conflict might be motivated by experiences they've had in different locations, so it is important to look beyond the local and to people's experiences and backgrounds in general.
- Look beyond who might be the most obvious stakeholders; even look to those who are technically not local residents and might not have legal rights to the resource in question.

—*The Editors*

References

Bahamas Department of Marine Resources. (2010). *Ministry of Agriculture and Fisheries: Bahamas Department of Marine Resources Five Year Sector Strategic Plan 2010–2014*. <http://bahamas.gov.bs>, accessed November 1, 2013.

Bahamas Department of Statistics. (2012). *Census of Population and Housing 2010 (Abaco and Exuma Cays)*. <http://statistics.bahamas.gov.bs>, accessed November 1, 2013.

Bahamas Environmental Health Services Act. (1987). *Environmental Health Services: Chapter 232*. <http://www.vertic.org/media/National%20Legislation/Bahamas/BS_Environmental_Health_Act_1987.pdf>, accessed March 21, 2015.

Bahamas National Trust. (2011). *The Bahamas Acts to Protect Sharks*. <http://www.bnt.bs/_m1840/press-releases/The-Bahamas-Acts-to-Protect-Sharks>, accessed November 1, 2013.

Bahamas National Trust. (2013). *Exuma Cays Land and Sea Park*. <http://www.bnt.bs/_m1731/The-National-Parks-of-The-Bahamas/Exuma-Cays-Land-and-Sea-Park>, accessed November 1, 2013.

Bojorquez-Tapia, L. A., de la Cueva, H., Diaz, S., et al. (2004). Environmental conflicts and nature reserves: redesigning Sierra San Pedro Martir National Park, Mexico. *Biological Conservation*, **117**(2):111–26.

Broad, K. and Sanchirico, J. N. (2008). Local perspectives on marine reserve creation in the Bahamas. *Ocean and Coastal Management*, **51**(11):763–71.

Chess, C. and Purcell, K. (1999). Public participation and the environment—do we know what works? *Environmental Science and Technology*, **33**(16):2685–92.

Clauzel, S. (2011). *Case Study of the Bahamas (Exuma) Demonstration Project: Marine Waste Management at Elizabeth Harbour, Exuma, Bahamas*. <http://iwcam.org/documents/gef-iwcam-project-knowledge-documents/gef-iwcam-demonstration-project-case-studies/case-study-of-the-gef-iwcam-bahamas-exuma-demonstration-project/view>, accessed November 1, 2013.

Coulthard, S., Johnson, D., and McGregor, J. A. (2011). Poverty, sustainability and human wellbeing: a social wellbeing approach to the global fisheries crisis. *Global Environmental Change*, **21**(2):453–63.

Craton, M. and Saunders, G. (1998). *Islands in the Stream: A History of the Bahamian People. Vol. 2*. Athens, GA: University of Georgia Press.

Evely, A. C., Pinard, M., Reed, M. S., and Fazey, I. (2011). High levels of participation in conservation projects enhance learning. *Conservation Letters*, **4**(2):116–26.

Government of The Bahamas. (2011). *About the Department of Marine Resources*. <https://www.bahamas.gov.bs/>, accessed November 1, 2013.

HWCC. (2008). *Benefitting Conservation Through Conflict Transformation White Paper, Washington, DC*.

IWCAM. (2013). *Exuma Project Document: Wastewater Management at Elizabeth Harbour Marina*. <http://iwcam.org/>, accessed November 1, 2013.

Keen, M. and Mahanty, S. (2006). Learning in sustainable natural resource management: challenges and opportunities in the Pacific. *Society and Natural Resources*, **19**(6):497–513.

Knapp, C. R., Iverson, J. B., Buckner, S. D., and Cant, S. V. (2011). Conservation of amphibians and reptiles in the Bahamas. In Hailey, A., Wilson, B. S., and Horrocks, J. A. (eds), *Conservation of Caribbean Island Herpetofaunas*, Vol. 2. Leiden: Brill, pp. 53–87.

Lederach, J. P., Neufeldt, R., and Culbertson, H. (2007). *Reflective Peace Building: A Planning, Monitoring and Learning Toolkit*. Notre Dame, IN: The Joan B. Kroc Institute for International Peace Studies, University of Notre Dame and Catholic Relief Services.

Luyet, V. (2005). *A Framework for the Participative Process in a Large Environmental Project. Case Study: The 3rd Rhone River Correction*. PhD thesis. Lausanne: Swiss Federal Institute of Technology.

Luyet, V., Schlaepfer, R., Parlange, M, and Buttler, A. (2012). A framework to implement stakeholder participation in environmental projects. *Environmental Management*, **111**:213–19.

MacGregor, J. (2000). *Exuma Coastal Zone Management Plan*. Prepared by ecoplan:net for the Bahamas Ministry of Tourism.

Madden, F. (2011). *Human–Wildlife Conflict Collaboration, Conservation Conflict Resolution Training Workshop*. Front Royal, VA: Smithsonian–Mason School of Conservation.

Madden, F. and McQuinn, B. (2014). Conservation's blind spot: the case for conflict transformation in wildlife conservation. *Biological Conservation*, **178**:97–106.

Martin, A. and Sherington, J. (1997). Participatory research methods: implementation, effectiveness and institutional context. *Agricultural Systems* **55**(2):195–216.

Mascia, M. B. (1999). Governance of marine protected areas in the wider Caribbean: preliminary results of an international mail survey. *Coastal Management*, **27**(4):391–402.

Maycock, D. (2010). *Size Matters Campaign Critical Analysis. Rare Conservation Organization*. <http://rareplanet.org>, accessed November 1, 2013.

Redpath, S. M., Young, J., Evely, A., et al. (2013). Understanding and managing conservation conflicts. *Trends in Ecology and Evolution*, **28**(2):100–09.

Reed, M. S. (2008). Stakeholder participation for environmental management: a literature review. *Biological Conservation*, **141**(10):2417–31.

Schusler, T. M., Decker, D. J., and Pfeffer, M. J. (2003). Social learning for collaborative natural resource management. *Society and Natural Resources*, **16**(4):309–26.

Stringer, L. C., Prell, C., Reed, M. S., et al. (2006). Unpacking "participation" in the adaptive management of socio-ecological systems: a critical review. *Ecology and Society*, **11**(2):39.

Sullivan-Sealey, K. (2000). The environmental impact of tourism: a study of Elizabeth Harbour. *Bahamas Journal of Science*, **7**, 2–19.

Thompson, P. R., Wailey, T., and Lummis, T. (1983). *Living the Fishing*. London: Routledge & Kegan Paul.

Young, J. C., Marzano, M., White, R. M., et al. (2010). The emergence of biodiversity conflicts from biodiversity impacts: characteristics and management strategies. *Biodiversity and Conservation*, **19**(14):3973–90.

UNEP-CAR/RCU. (2014). *Case Studies, Lessons Learnt, and Recommendations in the Development, Implementation and Management of GEF Projects in the Wider Caribbean Region (WCR)*. United Nations Environment Programme CEP Technical Report 59.

3

Transforming Wicked Environmental Problems in the Government Arena

A Case Study of the Effects of Marine Sound on Marine Mammals

Jill Lewandowski

3.1 Introduction

Government management of marine resources is wrought with highly complex and controversial issues—what have been called wicked environmental problems or intractable issues (Ludwig et al., 2001; Kreuter et al., 2004; Balint et al., 2011). Wicked environmental problems are characterized as issues having high levels of scientific uncertainty on risks, intermingling political/regulatory complexities, regularly evolving ecological and social environments, and diverse stakeholder values and viewpoints. Given this mix of challenges, productive decision-making is difficult at best (Rittel and Weber, 1973; Turnpenny et al., 2009). Couple this with traditional federal linear decision-making approaches, and effective marine resource management can be near impossible (Balint et al., 2011).

This chapter will undertake a case study of a current wicked environmental problem: U.S. federal management of anthropogenic noise and its effects on marine mammals (Figure 3.1). This chapter will show why this issue has become intractable and the challenges imposed under current linear federal decision-making approaches. It will describe an alternative framework, built on conflict transformation, that brings diverse interests (technical and non-technical) to the table to collaborate, create trade-offs and synergies, and ultimately foster effective action.

3.2 What makes an issue wicked?

Wicked environmental problems, or seemingly intractable conservation issues, share common characteristics, including (1) a high level of scientific uncertainty, (2) political and regulatory complexity, (3) diversified interested party interests, (4) a history of conflict and

Human–Wildlife Conflict: Complexity in the Marine Environment. Edited by Megan M. Draheim, Francine Madden, Julie-Beth McCarthy, and E. C. M. Parsons © Oxford University Press 2015. Published 2015 by Oxford University Press.

Figure 3.1 Large commercial ships passing through important feeding grounds of the North Atlantic right whale. Photo credit: Kate Sardi for the Stellwagen Bank National Marine Sanctuary/NOAA

resulting distrustful relationships, and (5) decision-making approaches that only increase conflict and intractability. Scientific uncertainty leads to many unknowns regarding the risks of decision options under consideration. Diversified interested party perspectives (influenced heavily by individual values) lead to disagreements on the problem definition and the correct path forward. Political complexities and a regularly evolving ecological and social environment further complicate solution building. Decision processes are often too simplified for the complexity of the issue, and this only lends to increased conflict. It is the mix of these characteristics that leads to an issue becoming wicked or intractable and remaining so (Chapter 1; Kreuter et al., 2004; Balint et al., 2011; Madden and McQuinn, 2014).

Another hallmark of a wicked environmental problem is that the conflict between some, if not all, of the interested parties is identity based. Identity conflict is where parties make assumptions and hold prejudices about others based on their group affiliation (Madden and McQuinn, 2014). Trust is low among these parties. People assume that an individual from another group will act or think a certain way and there is little hope for change (Madden and McQuinn, 2014). This is further reinforced by the fact that groups have actually established firm public positions as a response to years of conflict and frustration. Such a cycle of conflict will continue unless actions are taken to build the capacity of stakeholders to see past the established positions and affiliations, learn to communicate openly, and uncover common ground where it may exist (Fisher et al., 1991; Lederach, 2003; Madden and McQuinn, 2014).

Transforming this conflict out of the identity-based level is the central and key step to taming the wickedness of the issue.

3.2.1 The case study of the effects of anthropogenic sound on marine mammals

The federal management of the effects of anthropogenic sound on marine mammals is a good example of a marine conservation issue gone wicked. Anthropogenic sound, or human-made sound, in the ocean environment is an integral part of many human activities critical in supporting continued U.S. economic and social welfare; examples of such activities include vessel operation for commercial fisheries and the transport of goods/services, exploration and production of both traditional (e.g., oil and gas) and renewable (e.g., wind and tidal power) energy sources, exercises for military preparedness and national defense, dredging of offshore sand for beach and barrier island improvements (hurricane protection), seismic research for earthquake detection, and even recreational boating (e.g., nature tours, fishing trips, and weekend boaters; Richardson et al., 1995; Nowacek et al., 2007; Southall et al., 2007; Weilgart, 2007; OSPAR, 2009). From the perspective of the biological environment, however, anthropogenic sound can equal noise pollution.

3.2.1.1 *High level of scientific uncertainty*

Science shows that marine mammals produce and use sound to communicate as well as to orient, to locate and capture prey, and to detect and avoid predators (Richardson et al., 1995; Southall et al., 2007). When anthropogenic noise occurs within marine mammal hearing ranges and is at a high enough intensity, research has shown that exposures can in some instances lead to adverse physical and psychological effects on marine mammals. Possible effects can include (1) permanent or temporary hearing loss, discomfort, and injury; (2) masking of important sound signals; (3) behavioral responses such as fright, avoidance, and changes in physical or vocal behavior; and (4) indirectly altering prey availability (Nowacek et al., 2007; Southall et al., 2007; Clark et al., 2009).

Decades of research have largely answered some important questions, such as what are likely situations where sound may cause hearing damage or direct mortality, and what measures should be taken to avoid these situations. However, many key questions remain unanswered. Scientific results may also answer one question but raise many more in the process. For example, there is still scientific uncertainty regarding the nature and magnitude of behavioral impacts and whether these impacts may go beyond the individual animal and result in population-level effects. There are also unknowns about how the cumulative effect of multiple noise sources (e.g., shipping plus seismic) may affect animals.

3.2.1.2 *Political and regulatory complexity*

Many countries have laws in place for the protection of marine mammals. These same countries also have laws that promote resource development and related ocean uses (e.g., offshore energy development, mining, commercial shipping, fisheries, and military preparedness exercises). It is unclear how these various statutes relate to each other and whether the goals of one statute trump the goals of another.

These regulatory challenges are further compounded by political realities. For the most part, the industries being regulated on this issue are perceived as considerably large and influential (i.e., they use effective lobbying). They understandably want a reasonable decision in a timely manner and will exert their political influences when needed. Politicians also need to weigh the advancement of certain national issues (e.g., increased domestic energy production or military readiness) within the context of their own political reality. Environmental organizations launch campaigns to capture the public's interest on an issue that is otherwise largely unknown. They may even litigate particular decisions or activities, and this threat of litigation can also influence a decision. Ultimately, however, this mix of regulatory uncertainty, lobbying, litigation, and political and media campaigns bind government agency staff time, result in agency indecision (or unsustainable decisions), and add to the depth of the conflict among parties.

3.2.1.3 *Diversified interested party interests*

The diversity of stakeholder groups and the highly technical nature of this issue make for very complex relationships between parties. Stakeholders have diverse backgrounds, philosophies, and expertise. They include many governing bodies, industry groups, academic and research institutions, militaries, contractors, and environmental organizations. Examples of the stakeholder groups related to this case study are provided below (listed alphabetically) as well as a high-level view of the *perceived* object or motivating factor of each group (emphasis is added here on *perceived*, given that these factors are based on trends identified in a literature review alone and have not yet been ground-truthed directly with members of these groups, although such an analysis is currently taking place (Lewandowski, in progress)). Although this list is U.S.-centric, many of these group types are also mirrored in other countries.

- Academics: Some academic and research institutions conduct research that incidentally produces sound, such as seismic studies for earthquake preparedness (geo academics). Others are scientists who study the impact of sound on marine mammals (bio academics). All generally seek to reduce scientific uncertainty as a means to better inform decision-making, but this is where most commonalities end. Geo academics need to introduce sound into the marine environment as a means to gather data. Bio academics are focused on studying the effects of sound on marine life. All seek to increase scientific knowledge, but at times the objectives of each group can be at odds.
- Government: In the United States, approximately ten federal agencies are engaged on this issue. Agencies are charged with instituting their statutory mandate. Often, these mandates can dramatically differ between agencies and even be directly opposed to each other. For example, one agency may be charged with protecting marine mammals while another is mandated with developing ocean resources (albeit in an environmentally responsible manner). There is no overarching guidance on which mandate takes priority when conflict occurs.

- Contractors come in a variety of forms and serve nearly all of the other stakeholder groups. They bring outside expertise to help in drafting environmental reviews, develop and run modeling, and provide personnel to implement required mitigation measures (e.g., protected species observers). Overall, as small businesses, their main goals are to sustain their business, but they also add to the knowledge base and assist clients in meeting stated objectives.
- eNGOs: An increasing number of eNGOs are now engaged on marine sound issues, at both the local and national levels. Some appear either outright opposed to certain human activities (e.g., oil and gas development or military sonar) while others appear to seek a better balance between these activities and environmental protection. All aim to represent their membership (largely comprised of members of the public) in working to protect marine mammals from unnecessary harm.
- Industry: Ultimately, industries are focused on providing services and economic gain. This does not necessarily mean that industries are unequivocally seeking monetary gain at the expense of the environment. In fact, many industries and even individual companies are willing to adjust (i.e., mitigate) their activities in ways that better protect the environment but want these adjustments to be reasonable, effective, and proven, especially given the high cost of some mitigations. In addition, companies need timely government decisions in order to align with their business planning. Predictability is very important. However, the timing of government decisions is often out of sync with the longer-term planning processes typical of many industries.
- Military/navy (also federal agencies): In the United States, the navy conducts activities that are meant to prepare and defend the nation. Some types of naval sonar have been shown to result in injury and mortality to certain marine mammal species in very specific situations. In response, the navy developed a robust research program to better inform decisions and also implemented mitigations to reduce potentially harmful impacts.
- Tribal governments: Many tribes have historic land connections to coastal areas and long-standing cultural connections to marine mammals. In the United States, these tribes have inherent rights to self-government and are considered sovereign nations. Some of these tribes have engaged on ocean-sound issues, whether from an intrinsic concern of impacts on marine mammals (as shown by, e.g., native Hawaiians) or a more utilitarian concern over the availability of marine mammals for subsistence purposes (as shown by, e.g., Makah, and Alaska natives).

3.2.1.4 What are the existing levels of conflict?

Madden and McQuinn (2014) describe a useful tool called the levels of conflict model for analyzing the conflict surrounding an issue (see Chapter 1). Because of this issue's history and the relationships between major actors, most of the conflict today lies at the identity level. Interested parties are now largely entrenched in their positions, and their interpretations of or assumptions about the positions of others. Further, the processes meant to find resolution over the years have been more suitable for dispute level conflicts. This has resulted in short-sighted decisions, often leaving parties unsatisfied. Instead, new approaches are needed that focus on building capacity among stakeholders so that the underlying and identity conflicts can be transformed into more effective action for all parties.

Dispute level

According to Madden and McQuinn (2014), dispute level conflict represents the more immediate or point-in-time disagreement. The conflict will be settled at the dispute level if parties feel satisfied with the process, relationships, and the decision made. If any element is unsatisfactory, then the scarring from one specific dispute turns into underlying conflict that resurfaces quickly when the next dispute arises. (See also Chapter 1 for additional description of dispute level conflict.)

Early on, conflict on this issue was at the dispute level, and disagreement was on specific projects. There was little past history among stakeholders, and concerns about noise were just arising among a few groups.

One of the first major disputes about ocean noise and marine mammals involved the acoustic thermometry of ocean climate (ATOC) experiment. In 1994, a consortium of 11 research institutions in seven nations, with funding from the U.S. Navy, proposed the ATOC experiment (see <http://atoc.ucsd.edu>). The study's goal was to assess climate change by using low-frequency noise to determine changes in water temperature over time. Some academics were concerned about effects from this study on marine mammals, and several eNGOs launched media and membership campaigns to raise public attention. In response, the applicants proposed a marine mammal monitoring program (run by a third-party neutral university). The project was ultimately approved and ran from 1996–2006. Final third-party monitoring results indicated no significant or long-term effect to marine mammals, although questions were raised about the ability to effectively monitor for effects. At the end, several parties remained largely unsatisfied with the outcomes of this process, and this laid the foundation for underlying conflict in future issues.

Underlying level

Underlying conflict represents a history of unresolved disputes that continue to influence present-day interactions (Madden and McQuinn, 2014). Individuals are more likely to bring frustrations from past conflicts to the table regardless of the specific issue at hand. It becomes harder and harder to find common interests when the atmosphere is colored with past conflict. (See also Chapter 1 for additional description of underlying level of conflict.)

The issue of ocean sound and effects on marine mammals has been ongoing for at least two decades. Many of the key stakeholder groups on this issue, and even individuals within these groups, have been involved for years. Some oppose activities, regardless of the location and project type, as a means to draw and hold a line in the sand against fossil-fuel expansion or increased military activities. Often, the search for common ground has been put aside for established public positions, given the longer history of unresolved disagreements. Among some key parties, there is little trust and mostly disagreement. Because the commonly used public-policy processes for this issue are not built to openly address conflict, and some even seek to avoid it, frustration levels and ultimately conflict continue to increase.

Examples of underlying conflict on this issue are abundant. Two key ones are noted below. Again, these examples are meant to provide a high-level view of the *perceived* conflict. (Emphasis is added here on *perceived*, given that these factors are based on trends identified in a literature review alone and not yet ground-truthed directly with members of these groups, although such an analysis is currently taking place (Lewandowski, in progress).) Further, these examples are a culmination of past disputes.

Navy sonar: The navy and several eNGOs have been in litigation for the better part of eight years on the navy's use of active sonar systems for submarine detection (Zirbel et al., 2011). The fight has even resulted in a court case reaching the U.S. Supreme Court (*Winter v. Natural Resources Defense Council*, 2008). The issue of how to balance military preparedness and national security with protection of marine mammals from sound has been a central conflict in this issue.

The ATOC dispute, described in the previous section, in some ways was the start of the conflict between the U.S. Navy and certain eNGOs. Although ATOC was primarily an oceanographic experiment by oceanographic institutions, some funds and resources used were supplied by the navy. The experiences of both parties during this dispute (e.g., the navy feeling that eNGOs did not understand science and were unnecessarily inflaming the issue, and the eNGOs feeling that the navy and others should have been more forthcoming about the experiment) were carried over into disputes to come, first in the mid-1990s over the navy's use of low-frequency active sonar systems (LFA) and later over its use of mid-frequency active sonar systems (MFA).

Initially, the conflict over the navy's use of LFA was addressed through concerted efforts to dialog, identify issues, and attempts to address concerns. However, struggles arose that ultimately led to the end of dialog and extensive litigation. For example, the navy was unable to be completely transparent on operation of the LFA system due to national security concerns. Further, the navy was out of compliance with several environmental regulatory requirements over the use of LFA. Both led to an increased level of distrust on the part of the involved eNGOs. The navy, in turn, grew frustrated with a perceived lack of understanding by the eNGOs, of both the technology and the potential biological effects to marine mammals. In addition, the navy grew more distrustful as the involved eNGOs published public information about LFA and its impacts that was incorrect and also increased their media and public campaigns about the issue with inflammatory language that villainized the navy. The underlying conflict with LFA sonar ultimately set the stage for the debate to come over the navy's use of its MFA sonar—a debate mired in identity conflict and described in more detail later in this chapter.

Research seismic surveys: In 2012, the Pacific Gas and Electric (PG&E) requested approval from both federal and state regulators to conduct seismic surveying using airguns to assess earthquake risk at the Diablo Canyon nuclear reactor in San Luis, California. The project was initiated to check safety concerns, given the Fukushima Daiichi nuclear disaster following the 2011 earthquake and resulting tsunami in Japan. The project was a partnership with the U.S. National Science Foundation (NSF), who was to lead the surveys, given its mission of oceano-graphic and offshore earthquake probability research and its vessel and seismic surveying capabilities.

At the surface, public safety issues would seem to outweigh environmental concerns. However, existing underlying conflict was at play. The involved eNGOs came to the project with years of frustration over what they perceived to be an unsatisfactory resolution of the larger community addressing potential impacts of seismic surveys on marine mammals (whether by the NSF or by oil and gas exploration companies). The public already held nega-tive opinions over the management of the Diablo Canyon facility. There were also general antinuclear power sentiments. Several eNGOs launched media and membership campaigns against the proposed survey, stating the surveys were not worth the risk to marine mammals. Through this outreach, fishers entered the debate with concerns about noise reducing fish catches.

In reaction, PG&E and NSF reduced the survey size and increased protective measures. In their minds, they were accommodating each concern as it was raised to the best of their abil-ity. They became increasingly frustrated with eNGO public campaigns and felt that eNGOs were purposefully spreading misinformation to raise more opposition to the project and also increase fundraising. They also became even more aggravated over a perceived inefficient handling of the project by the federal regulators, a feeling that existed between NSF and regu-lators prior to this project.

Ultimately, the permit was denied by the state regulator, the California Coastal Commis-sion, which prevented any federal approval of the project (see <http://documents.coastal. ca.gov/reports/2012/11/W13b-11-2012.pdf>). The underlying conflict prevented resolution on this project. Tactics used by all sides further inflamed the issue. All parties left with a greater level of distrust and more embedded perceptions about each other that may further compli-cate the ability of all parties to work toward an acceptable solution on future projects.

Identity level

Identity conflict occurs where parties make assumptions and hold prejudices about others based on their group affiliation (Madden and McQuinn, 2014). Trust is often at an all-time low. This in turn makes resolving conflict even more challenging, given the assuming person has little to no hope that an individual from another group will act or think differently than expected (Madden and McQuinn, 2014; see also Chapter 1 for an additional description of identity-level conflict).

Identity conflict increases the intractability of the problem and feeds distrust. This results in a lack of meaningful dialog, and indecision or bad decisions. Stakeholders become seg-mented into group positions, with each group developing separate approaches to improve the situation from their point of view. Given deeply held assumptions about other groups, identity conflict is self-reinforcing in that these prejudices lead to avoidance of working together, which, in turn, inhibits collaborating toward a common vision of improving outcomes for all. The issue only becomes more wicked.

In the case of marine sound, underlying conflict has festered over the years and, in some cases, has deepened into identity-level conflict. Many, but not all, of the interested parties view this as "us" versus "them." While certain individuals can see past this and are able to "separate the people from the problem," the majority of people still possess long-held assumptions and prejudices about how individuals will behave according to their group affiliation.

Perhaps one of the best examples in marine sound of identity-level conflict is between the involved eNGOs and the navy. From the eNGO perspective, their mission is to promote sustainability and protect environmental resources. They are the voice for the public, and the watchdog on government activities. Achieving their mission is challenging when they are faced with perceived power inequalities with the navy, which eNGOs feel has much greater numbers and resources. It is further challenging given the navy is a large organization, and there is no one person, or even small group of people, to work with to effect change.

In order to attempt to balance the power, eNGOs have undertaken several main tactics. First, they built an alliance, with one organization identified as the lead, in order to share resources and coordinate efforts. Second, eNGOs, including this lead organization, have used litigation on multiple occasions where they felt dialog was ineffective. Third, the eNGOs have also conducted media and public campaigns to raise awareness of the issue and also fund their efforts. However, to gain media and public attention, eNGOs have needed to dramatize the issue and highlight worst-case scenarios.

The navy's mission is to protect national security and conduct military preparedness exercises. In order to do so, the navy needs to be able to continue training and wartime operations, activities which require a level of predictability and certainty for planning purposes. Some tactics used by eNGOs directly oppose this, such as litigation or media and public campaigns that bring attention to projects and drive public will to oppose them. This can lead to last-minute stoppage of exercises and ultimately a feeling of a lack of predictability and empowerment. The navy has undertaken several tactics to address this power struggle, including reducing transparency (protecting information) and even seeking a national defense exemption from Congress under the U.S. Marine Mammal Protection Act for MFA sonar exercises. The navy has also employed a tactic of increasing scientific understanding and detection of impacts through the development of a robust research and monitoring program.

Ultimately, the tactics of each side in their attempts to gain or equalize power work to further entrench the conflict and become the identity of each group to the other. The eNGOs use of exaggerated and/or inflammatory language in their public and media campaigns is often seen by the navy as dishonest and unprofessional, especially given the navy's internal culture, which promotes cordiality, respectful conversation, and the sharing of accurate information. Over time, the eNGOs' use of litigation has gained them substantial power, especially in terms of garnering the attention of federal regulators and eventually slowing regulatory decision-making (and hence approval of some naval operations). In turn, the navy now believes eNGOs use litigation as only a means to stop or slow activity and ultimately be anti-military. As a result, the navy has become much more careful about information it shares and limits any dialog with eNGOs due to beliefs that eNGOs will use this information against them in the court of law and/or public opinion. The eNGOs then feel even more excluded and denied access to information, and the cycle continues.

The development of this conflict over time to the identity level has also resulted in challenges for an individual from either side to reach across the divide. For example, the eNGO alliance (based on an overarching fear of their environmental identity being threatened) can inadvertently limit the diversity of perspectives and options for a path forward. The assignment of a lead eNGO organization means the interaction is mainly limited to that organization, especially for a technical issue like marine sound, for which maintaining an understanding of the issue requires regular involvement. To keep the eNGO alliance strong and hold the line with the navy, each eNGO must keep the same general public position and tactics as the lead organization, even if they individually may be open to more options and different approaches. The lead organization can also potentially become trapped in this alliance, as it may become hard for them to explain a change in tactic (e.g., moving to a position of compromise) to other organizations that could see this change as a threat to the environmental identity.

Reaching across the divide is now also difficult for individuals within the navy, where compromise and sharing of information works against their own identity. Further, participant turnover is high, given that management positions within the navy are term limited and people rotate through every two to three years. Whereas this turnover would seemingly lead to more diverse opinion, it can actually limit the time someone has to understand the issue enough during their term to move past engrained perceptions and long-held stances to alternative solutions. While two to three years may seem sufficient, it is not so in a large organization like the navy, where many factions are engaged on the issue and none appear to hold dominion over the other.

Additional examples of some key identity-level conflicts include, but are not limited to, the following groups.

Academics: The identity of scientists is largely tied to his or her professional credibility. The need to maintain one's credibility, and thus reinforce one's identity, manifests itself in three ways. First, there may be reluctance for scientists to provide a "good-enough" answer or a solution too early in the problem. If they are wrong, then their credibility could be affected. Second, credibility can be lost to some if a scientist collaborates with and/or accepts money from noise producers as their research may be seen as tainted as a result. Third, credibility can also be impacted if a scientist engages in the public-policy process. There is still an element of the academic world that promotes scientific purity and identifies with the separation of science and policy. In these cases, an individual scientist who crosses over and advocates a certain policy approach may lose credibility. Scientists, therefore, end up less likely to engage in the public-policy process. Ultimately, a wedge is driven between the application of science to policy and weakens the ability of the all stakeholders to be aware of and understand the best available science in its decision-making.

eNGOs and oil/gas industry: eNGOs see the oil and gas industry as largely interested in the financial bottom line and having inadequate regard for the environment. They often consider science generated by industry as partial, invalid, and one sided, even if the project is led by a reputable and independent scientist. (This in turn makes some academics reluctant to work with industry on projects.) Often, eNGOs feel overwhelmed by the financial and personnel resources of companies. For example, eNGOs cannot match resources needed to attend meetings and even if they can attend, eNGOs cannot do it in numbers sufficient to match industry representatives. eNGOs may also feel overpowered by industry and their lobbyists regarding access to government agencies. In response, eNGOs have turned to public/media campaigns and litigation as a means to gain more power and force the implementation of protective measures they feel are supported by science and law.

On the other hand, oil and gas companies feel conflicts with eNGOs through litigation and public/media campaigns have resulted in the need to operate in an unpredictable environment—a highly undesirable state for any business. Companies point to numerous instances where proposed activities were significantly delayed or canceled due to eNGO pressure. At times, costly mitigations were imposed without a clear understanding of their need or effectiveness. Companies feel overpowered by the eNGOs use of media and public campaigns and the influence of these efforts as well as litigation on government regulators. They often feel that, based on past actions, eNGOs will never be satisfied and will litigate no matter what a company may do to negotiate for a balanced approach. This, in turn, makes it difficult for individuals within a company to advocate further compromise with eNGOs to their management and, in the business world, weakened senior level support makes any type of progress challenging and even unachievable (see Figure 3.2).

Federal government and others: Many stakeholders involved in this issue generally believe the government at large is overly bureaucratic, uninformed, and ineffective. Stakeholders are frustrated with the government's management of this issue, or at least the decision processes

Figure 3.2 Sperm whale beginning dive near oil production platform in the Gulf of Mexico. Photo credit: Christoph Richter for the Bureau of Ocean Energy Management

in place. Many see management practices as not being informed by the best science. Regulations are outdated, and decisions processes are insufficient. Some believe that government staff do not understand the science and/or are frustrated when they do not see decisions reflecting their own point of view of the best available science. Almost all feel that current laws inadequately address ocean-sound issues and wonder why better laws or regulations are not developed. While stakeholders acknowledge the individuals within government agencies engaged on this issue are truly trying, years of perceived inadequacies over time have led to a prejudicial assumption that the government is incompetent or incapable of handling this issue.

Government, on the other hand, feels everyone is a critic. Depending on the stakeholder perspective, mitigation and monitoring requirements are either too restrictive or not restrictive enough. Decisions occur too soon or too late. Agencies are constrained by established decision-making processes, legal requirements, and even political realities. Agencies seek meaningful input from outside parties but more often feel they have been at the receiving end of parties' advocating myopic positions or even criticizing decisions without providing input earlier in the public decision-making process. Given these myopic views, the government feels other stakeholders do not understand, or are not open to understanding, the implications of their decisions on the larger management needs as well as on other stakeholders. Over time, this has lent to government viewing many stakeholders as uninterested in seeking an achievable outcome (or good-enough approach) and centered instead on their own needs. Further, the high level of litigation on this issue by certain eNGOs has caused agencies to spend significant time and resources developing encyclopedic environmental analyses to protect themselves for fear of more litigation. The government now identifies these eNGOs as mainly interested in litigation and not collaboration; this perception, in turn, hampers dialog and trust between these two parties.

Tribal governments: Tribal governments want to protect their culture or identity, something that has clearly been at risk for hundreds of years. Conflicts arise out of concern or perceived imbalances between maintaining cultural practices and economic activities from outside interests. For example, on the North Slope of Alaska, tribal concerns focus on avoiding impacts on the availability of marine mammals for subsistence hunting. These tribes have been in conflict with industry and the federal government over plans to explore or develop oil and gas resources (although this conflict has progressed in a more positive manner in recent years). At times, tribes feel left out of decisions or that traditional knowledge is largely ignored. They are being increasingly affected by climate change, and the security of their culture and world has never been more at risk. This long-held sense of lack of power has resulted in identity-level conflict among these tribal governments.

3.3 How can government decision processes increase conflict?

In the United States, and likely many other areas in the world, the federal government has largely served as the nexus for all stakeholder groups on this issue. The government is where the overarching policy and individual permitting decisions are made. However, for a variety of reasons (mainly lack of staff and financial resources), the U.S. federal government has not yet built a decision-making process that can make this issue tamer.

There are five primary reasons why the current government decision processes cannot successfully address this case study (or any other seemingly intractable issue).

The decision process primarily uses a linear approach: The typical government decision process is linear, sometimes called the waterfall approach (e.g., gather data—analyze data—formulate solution—implement solution; Conklin, 2010; see Comment). A linear process oversimplifies a complex issue. It is often myopic, focusing on a particular aspect of the issue (e.g., permitting a specific application) and fragments the larger issue into many smaller pieces. There is then no overarching vision looking holistically at the issue. Complexity requires a greater need for study and analysis (National Research Council, 1996). Instead, wicked environmental problems require a decision-making process that is iterative, deliberative, adaptive, and collaborative (Balint et al., 2011). Only with such an alternative process can one overarching vision be created, implemented, revisited, and adjusted over time.

Comment

For another example of how a linear decision-making process can actually exacerbate conflict, see Chapter 7, which contains a discussion of how such a process has alienated many people around Hawaiian monk seal conservation.

—The Editors

Timing rather than quality becomes the essence of the decision: Agencies push forward to meet deadlines required by regulation or even political pressures. This time crunch forces the government to focus on the immediate need and not the longer-term strategy. Agency personnel are pressured to analyze available information in the allotted time, thus limiting time for more creative decision alternatives. This time crunch also pressures stakeholders to rush in influencing a decision. Instead of having enough time for a back-and-forth dialog, they must instead establish and hold strong "positions." Positions then create a competitive, adversarial, and distrustful environment with "opposing" parties and do little to help solve an intractable issue. Once people commit to positions, it can become part of their identity and therefore vital to defend (Fisher et al., 1991; Lederach, 2003; Madden and McQuinn, 2014). Ironically, this pressure to move forward under time constraints often results in longer time frames for decisions as agencies become knotted up with political and stakeholder pressures (e.g., responding to mass email campaigns, or briefing senior managers in preparation for meetings with stakeholders) and, in some cases, have to redo costly analyses due to litigation.

Process heavily emphasizes ecological science without addressing the role of human social dynamics: Many stakeholders feel that ecological science will provide the answer on intractable environmental issues, especially highly technical ones like marine mammals and ocean noise (see Box 3.1). Where multistakeholder workshops have occurred on this issue, they have largely focused on identifying scientific information needs, perhaps as this appears to be the most likely area of compromise among stakeholders. The government agency staff who develop the environmental analyses for permitting generally also come from science backgrounds. The legal framework requires the use of the best available information, largely interpreted to be scientific in nature. Agencies have lost litigation when not using the best available information. This ultimately leads agencies (and other stakeholders) to find more comfort in the science as a means to a solution (or for the agencies lessening the potential for successful litigation). Agencies often convene and rely on scientific experts to provide key input.

Box 3.1 Emphasis on ecological science in ocean noise and marine mammals

Out of the 230 major reports (workshops, opinion papers, and peer-reviewed reports) on ocean noise and marine mammals reviewed for this chapter, all focused on determining what available science said about impacts and what science and mitigating measures were needed to address the issue. This was consistent regardless of whether the report was generated by academics, government, industry, environmental NGOs, or even multistakeholder groups. None of these reviewed reports addressed the existing conflict among parties in any substantial manner, or how to openly consider the conflict in the context of improving outcomes on this issue.

There is no doubt that science will help inform and answer key questions and that increasing the understanding of an issue is a central component. However, the science cannot be pursued solely as the means for resolution. While the science is important, and should be pursued, this *expert-driven* approach comes at a cost. It largely ignores the issues of social values, equity, and justice that made the problem wicked or intractable to begin with. Rather, the process needs to recognize that political and social influences will overshadow any technical analyses on controversial issues (Renn et al., 1995). Agencies must instead develop new ways to use science *and* issues of social values and equity to inform decisions (Ludwig et al., 2001; Reed, 2008).

Parties are not truly part of the decision-making process: Environmental statutes promote transparency but do not go as far as requiring (or in some cases even allowing) stakeholders a seat at the table when making decisions. Where public input is allowed, the interaction is largely limited to written comments, timed oral comments at public meetings, and, in fewer cases, meetings between federal managers and stakeholders. Generally, the information flow is one-way from the stakeholders to the government, and agencies ultimately determine which comments they incorporate into their decisions. There is little opportunity for genuine, creative shared decision-making in the face of more reactive and arms-length notice and comment periods.

Participation in a well-designed decision process actually empowers parties to better understand all sides of the issue, have a greater opportunity to explain their viewpoints and listen to those of others, and consider the available information. Such involvement can promote trust that moves participants toward coordination and cooperation for mutual benefit—in both the short term and long term. This creates network power, where all participants share in the flow of power. With these networks come diversity, independence, and authentic dialog. This can then lead parties to open their minds to develop creative, workable solutions, compromise, and even accept decisions not fully aligned with their viewpoints (Fisher et al., 1991; Huer et al., 2007; Reed, 2008; Madden and McQuinn, 2014). It can also result in a network that is more capable of learning, adapting to change, and sustainable with the long-term vision generally needed to address wicked or intractable issues (Booher and Innes, 2002; Zhang and Dawes, 2006; Blackstock et al., 2007).

Importantly, participation of stakeholders in a decision-making process does not mean decisions need to be done by consensus or that the government has to relinquish control of a decision (since legally it is required to take action). Rather, for complex issues, processes can be designed within the parameters of regulatory requirements (including timing) that fully integrate parties as participants in the decision process. Along with this participation is an expectation that the final decision is at the discretion of the government but that the government will be open and transparent as to the reasoning behind its decision.

Process fails to address the heart of the issue—the need to transform conflict into effective action: Ultimately, intractable issues are a conflict between people about the environment (Madden and McQuinn, 2014). Decision-making processes cannot produce effective solutions in situations where conflicting goals, identities, and values predominate (Weber, 1985; Moote and McClaran, 1997; Conley and Moote, 2003; Madden and McQuinn, 2014). Addressing the conflict is therefore the most important action that can be taken and is necessary for the success of workable and sustainable solutions on intractable issues. It is also the most difficult to implement. Government management of marine mammals and ocean noise issues is bound to fail if this critical component of understanding and transforming conflict is missed (Madden and McQuinn, 2014; see Box 3.2). Although this may not be apparent at the short-term decision point, it will become apparent when agencies and stakeholders realize they keep revisiting the same set of issues.

Box 3.2 Failure of the 2004–2005 U.S. Federal Advisory Committee on Acoustic Impacts on Marine Mammals

During 2004–2005, the U.S. Marine Mammal Commission convened a Federal Advisory Committee (FAC) on Acoustic Impacts on Marine Mammals (Marine Mammal Commission, 2006). The FAC consisted of 28 nominated representatives. It met six times over a two-year period. The goals were to seek consensus on (1) evaluating available information, (2) identifying areas of general scientific agreement and uncertainty, (3) identifying research needs and priorities, and (4) recommending management actions to avoid and mitigate possible adverse effects (Marine Mammal Commission, 2006).

continued

Box 3.2 (*Continued*)

The end result of the FAC was a breakdown of stakeholder groups into various caucuses, each providing their own report, and very little to no overarching consensus (Marine Mammal Commission, 2006). What was meant to bring people together was largely seen as a failure. But why did it fail? Potential key reasons include but are not limited to:

- Striving for consensus set the bar too high for an issue with significant underlying and identity conflict. The group did not yet have the capacity to work together. Conflict was still very much in the way.
- FAC procedural requirements constrained flexibility and adaptability.
- Goals were science driven and used science as a means to resolution (e.g., what can the parties agree to in terms of scientific knowledge and needs?).
- Meetings were designed to focus on agreement and not fully address disagreement, a critical step if parties are to move off of hard positions, look toward common interests, and consider potential compromises.

The situation assessment conducted in advance of the FAC involved a third-party neutral interviewing 80 individuals (see U.S. Institute for Environmental Conflict Resolution, 2003). Questions asked mainly focused on science issues, reducing management controversy (but through quality risk assessments and appropriate application of mitigation), and factors for a successful collaborative process (such as a product focused approach, willingness to participate, and available scientific expertise). However, also within the situation assessment report were more illuminating or "indicator" statements such as deciding what battles needed to be fought and not creating unnecessary ones, improving relations among parties to make progress on the topic, balancing environmental concerns with economic and other concerns, promoting policies that respect and protect marine mammal welfare, identifying strained and personalized relationships among interested parties, and desiring a way forward that would put an end to what had been crises-driven, costly battles. As one interviewee stated, "It would be very easy, and not desirable, to get engrossed in the numerous scientific questions related to this topic and never get to the policy and management decisions that are the essence of the dissonance that we are currently experiencing" (U.S. Institute for Environmental Conflict Resolution, 2003). It is these statements that indicated a need for a process to transform the group into collective action.

3.4 How can a wicked issue be tamed?

> Try never to ignore or talk away someone's perception. Instead, try to understand where it is rooted.
>
> **J. P. Lederach (2003, p. 58)**

So, how does the government break the cycle and make an issue like anthropogenic noise and marine mammals less wicked or intractable? By turning away from traditional linear approaches and toward alternative approaches built on understanding the differing human values and identities that made the issue wicked to begin with, it can transform this conflict into more effective action (Lederach, 2003; Balint et al., 2011; Madden and McQuinn, 2014).

The difference in any alternate approach for a wicked problem is that resolution is not aimed at one point in time but rather at building (transforming) the capacity of involved stakeholder groups to work together in the longer term. Transformative processes focus more on the relationship among participants and seek to build change processes that address not only the immediate situation (short-term responsive) but also the broader setting creating the conflict (long-term strategic; Lederach, 2003; Madden and McQuinn, 2014).

Conflict changes relationships in predictable ways, altering communication patterns and social organization, and altering images of the self and of the other (Rupesinghe, 1994; Kriesberg, 1998; Botes, 2003; Lederach, 2003). A transformative process openly addresses perceptions of issues, actions, problem definition, and identity (of self and others) so that each group gains a more accurate understanding of the others (also called recognition by Bush and Folger (2004)). Understanding helps stakeholders to develop the capacity to see (look beyond the presenting problem) and empathy to understand the situation of another (Lederach, 2003). This transforms personal relationships that can then facilitate the transformation of the group social system (Dukes, 1999). Once this understanding is achieved, methods can be used to change the way the conflict is expressed and move the dialog from competition, or even aggression, to conciliation and cooperation. Participants are empowered to define their issues and seek their own solutions and can approach current and future problems with stronger, more open views (called empowerment by Bush and Folger (2004)). The conflict itself therefore becomes less destructive and less of a hindrance to making progress on potential paths forward (Schrock-Shenk and Ressler, 1999; Green, 2002). Through a transformative process, the group can then deal more effectively with future issues (Dukes, 1993).

3.4.1 Developing an alternative approach

Within this alternative approach of conflict transformation, agencies must undertake a rigorous analysis of the existing social conflict that goes far beyond the current situation assessment approach. Studies should be designed to uncover stakeholder values, interests, identities, established positions (and how they have changed over time), differences in problem definition, perceptions of self and other stakeholders, trust in the process (or how to build it if trust does not exist), and thoughts on better ways to design decision processes. This should entail robust qualitative research approaches (i.e., stakeholder interviews, observations of group interactions, and document reviews) that shed light on the context and reasoning behind the conflict rather than just preferences for policy choices, positions on issues, and what scientific studies people feel are needed. Further, the analysis must include interviews (rather than only written surveys), given that interviews are where the context and reasoning behind an individual's public stance can be revealed. Interviews afford people a full opportunity to explicitly and thoroughly state their perspectives in a one-on-one, private situation. Interviews can also provide anonymity and a level of comfort to share their individual (not stakeholder group-driven) perspectives and self-thinking (Berg, 2009; Angrosino, 2010). This is especially critical for intractable issues, where individual stakeholders are often welded to a publicly stated position, not the interests behind how they came to this position.

Agencies should then use these data to design and implement an iterative, deliberative, collaborative, and adaptive process built on transforming the conflict and growing group capacity to address issues in the long term. Chapter 1 in this volume provides key elements of a transformative approach. The design of an alternative approach should be done in concert with a conflict transformation expert, especially given that transformative processes are new to the government. It should also be done in collaboration with involved stakeholders and geared to the specific issue. There is no one size fits all for designing processes to address wicked problems.

Any transformative process will encourage stakeholders to be open and honest about the conflict. This is certainly new ground for the government, where conflict is held at arm's length, and the expression of emotions is generally perceived negatively. So often agencies convene groups of interested parties and, rather than address the conflict, forge ahead into where potential solutions (decisions) may lie. This is especially true on technical or scientific issues, such as marine mammals and noise, where stakeholders tend to have technical backgrounds and may be reluctant to move outside of factual discussions.

Stakeholders may believe that emotions make them weaker, interfere with good judgment and reasoning, and complicate planning. Modern neuroscience, however, proves that emotions actually make us more effective, are essential to good judgment, speed up reasoning, build trust and connection, and provide vital feedback. There is conclusive biological evidence that decision-making is neurologically impossible without being informed by emotions (Sanfey, 2007). Emotions are, in fact, highly intelligent and critical for building group intelligence and social capital (Kramer, 2007). Further, people need to be heard and allowed to express emotion before they can open their minds and consider compromise (Innes and Booher, 1999; Bush and Folger, 2004). They also need to understand the larger picture of the issue, including regulatory and political constraints, as well as how their positions affect other stakeholders. By allowing individuals to "emote," the conflict can be better understood by all participants and each can gain a greater understanding of the reasoning, beliefs, and potential areas of compromise (shared interests) with fellow stakeholders. It is then, and only then, that steps can be taken to transform the group and its conflict toward effective action, decisions, and outcomes.

3.4.2 Is it worth the time and cost?

An alternative approach will no doubt cost time and money, but so do failed linear processes, government indecisions (or poor decisions), and litigation (Ewel, 2001; Smith and McDonough, 2001; Irvin and Stansbury, 2004; U.S. Institute for Environmental Conflict Resolution, 2005; Agranoff, 2006). The costs of alternative processes should be considered as long-term investments, where benefits will ultimately outweigh costs. Benefits include social learning, gains in social capital, empowering groups to work together in the long term, and the probability of more effective and sustainable decisions. The costs of more traditional, linear processes generally include redoing lengthy and costly environmental analyses, missed opportunities for public and private investments from untimely decisions, deepening antagonism and hostility among stakeholders, and costly impacts to natural resources as protective actions are stymied by an inability to act on decisions. There is also a substantial cost from litigation associated with intractable issues. For example, a 2011 Government Accountability Office report found that the Department of Justice annually defends about 155 lawsuits against the U.S. Environmental Protection Agency alone (fiscal years 1995–2010). The costs borne by the government to defend these cases averaged $3.3 million annually for a total of $43 million (fiscal years 1998–2010). In addition, costs to settle cases (i.e., to avoid going to court) cost the government an additional $3.2 million annually (fiscal years 2006–2010; Government Accountability Office, 2011; all amounts are given in constant 2010 dollars).

There may also be a wariness of how such a process can fit into a regulatory environment where timely decisions are needed or where the government is mandated by law to make the decision. However, an alternative process can be designed within the directives and frameworks of the specific statutory requirements. Larger decisions can be broken down into phases with agreed-upon time frames. Participants can design a process that addresses short-term needs but also builds a longer-term vision. This collective effort, in turn, builds the capacity of the group to collaborate and develop more effective strategies for dealing with future problems and continuing cycles of conflict. In fact, such a process can actually *buy* time to reach a more effective overall decision, since interested parties are now part of working toward a solution, are more willing to accept short-term decisions if engaged in long-term solutions, and are much less likely to litigate in the interim (Tyler, 1998: Gangl, 2003; Heuer et al., 2007; Reed et al., 2008; Weber et al., 2008).

3.5 Conclusions

> The courts of this country should not be the places where resolution of disputes begins. They should be the places where the disputes end after alternative methods of resolving disputes have been considered and tried.
>
> **Justice Sandra Day O'Connor, former U.S. Supreme Court Justice**
> **(University of Pennsylvania, 2010)**

Ultimately, the wickedness of an issue is not about the science, nor will the science ever tame the issue on its own. Rather, the issue is intractable because of the conflict *between* people about the most appropriate path forward. It is then imperative to understand, address, and transform this conflict in order to move off the decision carousel (i.e., patterns of continuous and circular debates) and toward improved outcomes and sustainable decisions.

With the understanding of the role conflict plays in making an issue intractable will come the need to design new decision processes. This will require a paradigm shift in the government that moves away from shorter-term horizon, science-driven, linear processes to longer-term horizon, holistic, iterative, and adaptive approaches. Such a change will also require agencies and other stakeholders to step out of their "technical" comfort zones, address the conflict openly in a productive manner, and collaboratively develop the capacity to deal with both immediate and long-term aspects of the issue. Government should play a central role in initially guiding and growing this change, given government most often serves as a nexus for all other stakeholder groups.

It may seem too ambitious, costly or unrealistic to pursue such a change. Costs may seem too high and time too short. Stakeholders may be wary about openly addressing conflict or believe that certain groups are just unwilling to improve outcomes. However, most everyone will likely agree that the current approaches are not working, indecision or poor decisions occur, and the same set of issues are revisited again and again without significant progress.

As Albert Einstein once stated, "We cannot solve the problems we have created with the same thinking that created them." It is time to reset and test the role that transformative processes can play in truly taming wicked environmental problems.

Lessons learned

- Many policy decision-making processes have a very linear design. However, this can oversimplify complex conflict, which can make a satisfactory outcome harder to achieve.
- Participants should be aware of outside-imposed time deadlines. These can create pressure to reach a decision, pressure which can ultimately be counterproductive.
- Although good science should help to inform decisions, it can be counterproductive to have the science be the only focus of the decision-making process.
- Qualitative research helps to illuminate what is happening in cases of complex conflict. One-on-one anonymous interviews are key in truly understanding the interests (the "why") behind the positions (the "what") an individual public states are needed, and then developing new approaches to address the collective interests and find common ground.

—The Editors

References

Agranoff, R. (2006). Inside collaborative networks: ten lessons for public managers. *Public Administration Review*, **66**(1):56–65.

Angrosino, M. (2010). *How Do They Do That? The Process of Social Research*. Long Grove, IL: Waveland Press, Inc.

Balint, P., Stewart, R., Desai, A., and Walters, L. (2011). *Wicked Environmental Problems: Managing Uncertainty and Conflict*. Washington, DC: Island Press.

Berg, B. (2009). *Qualitative Research: Methods for the Social Sciences*, 7th edn. Boston, MA: Allyn & Bacon.

Blackstock, K., Kelly, G., and Horsey, B. (2007). Developing and applying a framework to evaluate participatory research for sustainability. *Ecological Economics*, **60**(4):726–42.

Booher, D. and Innes, J. (2002). Network power in collaborative planning. *Journal of Planning Education and Research*, **21**(3):221–36.

Botes, J. (2003). Conflict transformation: a debate over semantics or a crucial shift in the theory and practice of peace and conflict studies? *International Journal of Peace Studies*, **8**(2):1–27.

Bush, R. and Folger, J. (2004). *The Promise of Mediation: The Transformative Approach to Conflict*. San Francisco, CA: John Wiley & Sons, Inc.

Clark, C., Ellison, W., Southall, B., et al. (2009). Acoustic masking in marine ecosystems: intuitions, analysis, and implication. *Marine Ecology Program Series*, **395**:201–22.

Conklin, J. (2010). *Wicked Problems and Social Complexity*. Napa, CA: Cognexus Institute.

Conley, A. and Moote, M. (2003). Evaluating collaborative natural resource management. *Society and Natural Resources*, **16**(5):371–86.

Dukes, E. (1993). Public conflict resolution: a transformative approach. *Negotiation Journal* **9**(1):45–57.

Dukes, E. (1999). Why conflict transformation matters: three cases. *Peace and Conflict Studies*, **6**(2):47–66.

Ewel, K. C. (2001). Natural resource management: the need for interdisciplinary collaboration. *Ecosystems*, **4**(8):716–22.

Fisher, R., Ury, W., and Patton, B. (1991). *Getting to Yes: Negotiating Agreement Without Giving In*. Boston, MA: Houghton Mifflin Harcourt.

Gangl, A. (2003). Procedural justice theory and evaluations of the lawmaking process. *Political Behavior*, **25**(2):119–49.

Government Accountability Office. (2008). *Opportunities Exist to Enhance Federal Participation in Collaborative Efforts to Reduce Conflicts and Improve Natural Resource Conditions*. GAO-08-262. Washington, DC: Government Accountability Office.

Green, P. (2002). CONTACT: training a new generation of peacemakers. *Peace and Change*, **27**(1):97–105.

Heuer, L. Penrod, S., and Kattan, A. (2007). The role of societal benefits and fairness concerns among decision makers and decision recipients. *Law and Human Behavior*, **31**(6):573–610.

Innes, J. and Booher, D. (1999). Consensus building and complex adaptive systems: a framework for evaluating collaborative planning. *Journal of the American Planning Association*, **65**(4):412–23.

Irvin, R. and Stansbury, J. (2004). Citizen participation in decision making: is it worth the effort? *Public Administration Review*, **65**(1):55–65.

Kramer, R. (2007). How might action learning be used to develop the emotional intelligence and leadership capacity of public administrators? *Journal of Public Affairs Education*, **13**(2):205–30.

Kreuter, M., DeRose, C., Howze, E., and Baldwin, G. (2004). Understanding wicked problems: a key to advancing environmental health promotion. *Health Education and Behavior*, **31**(4): 429–40.

Kriesberg, L. (1998). Intractable conflicts. In Weiner, E. (ed.), *The Handbook of Interethnic Coexistence*. New York: Abraham Fund Publication, pp. 332–42.

Lederach, J. (2003). *The Little Book of Conflict Transformation: Clear Articulation of the Guiding Principles by a Pioneer in the Field*. Intercourse, PA: Good Books.

Ludwig, D., Mangel, M., and Haddad, B. (2001). Ecology, conservation and public policy. *Annual Review of Ecological Systems*, **32**:481–547.

Madden, F. and McQuinn, B. (2014). Conservation's blind spot: the case for conflict transformation in wildlife conservation. *Biological Conservation*, **178**:97–106.

Marine Mammal Commission. (2006). *Advisory Committee on Acoustic Impacts on Marine Mammals, Report to the Marine Mammal Commission*. <mmc.gov/reports/workshop/pdf/soundFACAreport.pdf>, accessed March 9, 2015.

Moote, M. and McClaran, M. (1997). Viewpoint: implications of participatory democracy for public land planning. *Journal of Range Management*, **50**(5):473–81.

National Research Council. (1996). *Understanding Risk: Informing Decisions in a Democratic Society*. Washington, DC: The National Academies Press.

Winter v. Natural Resources Defense Council, 555 U.S. 7 (2008).

Nowacek, D., Thorne, L., Johnston, D., and Tyack, P. (2007). Responses of cetaceans to anthropogenic noise. *Mammalian Review*, **37**(2):81–115.

OSPAR Commission. (2009). *Overview of the impacts of anthropogenic underwater sound in the marine environment*. *OSPAR Report No. 441*. <http://ospar.org/v_publications/download.asp?v1=p00441>, accessed March 9, 2015.

Reed, M. (2008). Stakeholder participation for environmental management: a literature review. *Biological Conservation*, **141**(10):2417–31.

Reed, M., Dougill, A., and Baker, T. (2008). Participatory indicator development: what can ecologists and local communities learn from each other? *Ecological Applications*, **18**(5):1253–69.

Renn, O., Webler, T., and Wiedemann, P. (1995). *Fairness and Competence in Citizen Participation: Evaluating Methods for Environmental Discourse*. Dordrecht: Kluwer Academic.

Richardson, W., Greene, C., Malme, C., and Thomson, D., eds. (1995). *Marine Mammals and Noise*. New York: Academic Press.

Rittel, H. and Webber, M. (1973). Dilemmas in a general theory of planning. *Policy Sciences*, **4**(2): 155–69.

Rupesinghe, K. (1994). *Protracted Conflict*. London: McMillan.

Sanfey, A. (2007). Decision neuroscience: new directions in studies of judgment and decision making. *Current Directions in Psychological Science*, **16**(3):151–5.

Schrock-Shenk, C. and Ressler, L., eds. (1999). *Making Peace with Conflict*. Harrisonburg, VA: Herald Press.

Smith, P. and McDonough, M. (2001). Beyond public participation: fairness in natural resource decision-making. *Society and Natural Resources*, **14**(3):239–49.

Southall, B., Bowles, A., Ellison, W., et al. (2007). Marine mammal noise exposure criteria: initial scientific recommendations. *Aquatic Mammals*, **33**(4):411–520.

Turnpenny, J., Lorenzoni, I., and Jones, M. (2009). Noisy and definitely not normal: responding to wicked issues in the environment, energy and health. *Environmental Science and Policy*, **12**(3): 347–58.

Tyler, T. (1988). What is procedural justice? Criteria used by citizens to assess the fairness of legal procedures. *Law and Society Review*, **22**(1):103–36.

University of Pennsylvania. (2010). Quote . . . unquote. *Almanac*, **57**(15):1.

U.S. Institute for Environmental Conflict Resolution. (2003). *Process Assessment Report for a Dialogue on Impacts of Anthropogenic Noise on Marine Mammals*. Washington, DC: U.S. Institute for Environmental Conflict Resolution.

U.S. Institute for Environmental Conflict Resolution. (2005). *Interagency Initiative to Foster Collaborative Problem Solving and Environmental Conflict Resolution: Briefing Report for Federal Department Leadership*. <http://www.udall.gov/documents/Institute/br.pdf>, accessed March 9, 2015.

Weber, M. (1985). A method of multiattribute decision making with incomplete information. *Management Science*, **31**(11):1365–71.

Weber, E. and Khademian, A. (2008). Wicked Problems, knowledge challenges, and collaborative capacity building in network settings. *Public Administration Review*, **68**(2):334–9.

Weilgart, L. (2007). The impacts of anthropogenic ocean noise on cetaceans and implications for management. *Canadian Journal of Zoology*, **85**(11):1091–116.

Zhang, J. and Dawes, S. (2006). Expectations and perceptions of benefits, barriers, and success in public sector knowledge networks. *Public Performance and Management Review*, **29**(4):433–66.

Zirbel, K., Balint, P., and Parsons, E. C. (2011). Navy sonar, cetaceans and the US Supreme Court: a review of cetacean mitigation and litigation in the US. *Marine Pollution Bulletin*, **63**(1–4):40–8.

4

Conservation in Conflict

An Overview of Humpback Whale (*Megaptera novaeangliae*) Management in Samaná, Dominican Republic

Christine Gleason

4.1 Introduction

In the late sixties, the scientific world at large was introduced to the phrase "tragedy of the commons." Hardin (1968) in his discussion of human population presented the idea that an open, unregulated, or common-pool resource would be utilized until its destruction. In Hardin's example, the over-utilization of a pasture, through the continued addition of livestock, created a greater benefit to the individual farmer per animal added. This farmer reaped all the benefits but only absorbed a portion of the negative impacts of overgrazing. The consequences of overgrazing were felt by all stakeholders.

The theory of the tragedy of the commons and common-pool resources has been applied to science and conservation in areas such as forest, ocean, pastureland, and wildlife tourism management (Carlsson and Berkes, 2006; Moore and Rodgers, 2010). In many of these cases, better technology or management schemes have been recommended to correct the negative effects of common-pool resources. However, Hardin himself argued that common-pool resources were in a class of problems that could not be saved by technology and natural sciences; rather, they required social change. In other words, to promote the conservation of the resource, we must deal with the stakeholders who utilize it, and the underlying conflict between them.

Managing conflict between stakeholders is a complex task. Many conflicts have a long and complicated past, and without the proper framework for transformation conservation, researchers and practitioners can make a tense situation worse. Madden and McQuinn (2014; also see Chapter 1) have adapted from the peace-building field an analytical model which allows researchers and practitioners to approach complex social and environmental conflict in a systematic, comprehensive, and holistic way. This model

Human–Wildlife Conflict: Complexity in the Marine Environment. Edited by Megan M. Draheim, Francine Madden, Julie-Beth McCarthy, and E. C. M. Parsons © Oxford University Press 2015. Published 2015 by Oxford University Press.

posits three levels of conflict: dispute, underlying, and identity based. These levels allow researchers to identify the outward conflict, or dispute, as well as go beneath the surface and analyze past unresolved conflicts and deeper-rooted identity-based conflicts. Through this analysis, the researcher or practitioner can uncover the deeper layers of conflict that may prevent the presenting dispute from being settled. By using this framework and understanding the principles and skills of conflict transformation, researchers and practitioners can help facilitate change. By including conflict transformation as a part of conservation initiatives, a more inclusive and sustainable management plan can be created.

This chapter will examine the conflict within the co-management system developed for regulating whale-watching in the Dominican Republic. Co-management, as defined by Berkes et al. (1991, p. 12), is "the sharing of power and responsibility between the government and local resource users." It assumes an equal sharing of power between stakeholders; however, this is rarely the case. Constant monitoring of the co-management system is needed to ensure an equal division of labor and stakeholder control. In order for co-management to be effective, an adaptive management approach that incorporates a complete understanding of and capacity to address the complex social dynamics influencing co-management efforts must be employed to ensure all stakeholder issues are addressed. Without it, conflict may impede success. Unfortunately, most evaluations of whale-watching management systems focus on compliance with current guidelines. Few evaluate whale-watching through the lens of common-pool resources and the social conflicts that can occur within them (Lawrence et al., 1999; Neves-Graca, 2004). This is disconcerting, as whale-watching companies primarily operate in coastal areas with highly concentrated human and cetacean populations. Creating a management system that benefits both people and wildlife is challenging (Weinstein and Reed, 2005), as the needs and uses of each can create incompatibilities that may create the conditions for conflict. The history of how this management system was set up, and current or historical conditions and conflicts present within the larger system, will give indications as to what type of process is needed to ensure effective and sustainable management.

Defining what conflicts exist within a system is a critical step to conflict transformation (Adams et al., 2003; Madden and McQuinn, 2014). Often, stakeholders have different social, economic, and knowledge backgrounds, as well as different responses and receptivity to a given situation. This diversity creates varying opinions and perceptions within a community. In many cases, stakeholders do not realize how persuasive their knowledge and experiences are and how they may be influencing current conflicts (Adams et al., 2003). This chapter will provide an overview of the varying viewpoints which create conflict within the co-management system in the Dominican Republic. Furthermore, it will explore how outside entities and social norms influence current social conflict. Specifically, it will review how local tour operators, international tour companies, and cruise ships play a role in the compliance levels of whale-watching guidelines and therefore conflict within the system. It will serve only as an analysis of conflict, as specific recommendations as to how to transform this conflict are outside the scope of this chapter.

4.2 A brief history of tourism in the Dominican Republic

Located on the island of Hispaniola (shared with Haiti), the Dominican Republic (Figure 4.1) is a Spanish-speaking country with over 1,200 km of coastline (Fuller, 1999; Draheim et al., 2010). It has an approximate population of ten million people, with roughly 40% of those citizens living in poverty (World Bank, 2012). Originally colonized by Spain, which was followed closely by France and Haiti, the Dominican Republic won its independence in 1844. Its history of occupation and political intervention by the United States, however, has helped to mold current Dominican culture and tourism (Leon, 2004).

International tourists began arriving in the Dominican Republic in 1967 because of Presidential Decree #2536–1968, which made tourism a national priority. The government hired foreign consultants to help encourage private investments and gave tax breaks to foreign and domestic tourism partners (Fuller, 1999). Tourism grew rapidly, and it became evident that its unrestricted growth was having negative impacts. Although government incentives were phased out, today this history of foreign investment can still be seen (Fuller, 1999; Leon, 2004).

Although originally domestically owned, today the vast majority of the tourism industry is foreign owned. This is primarily due to the large amount of currency needed to make a successful investment (Tejada, 1996; Fuller, 1999). In 2003, 70% of hotel rooms in the Dominican Republic were located in foreign-owned resorts (ASONAHORES, 2003). This foreign-dominated tourism trade leads to social conflict over tourism resources such as beaches and local wildlife. In fact, as large resorts create private beaches for their cliental, locals are shut out of beaches they have been visiting for generations (Gleason, unpublished observations). Even if locals are not shut out completely, there are often segregated local and tourist sections

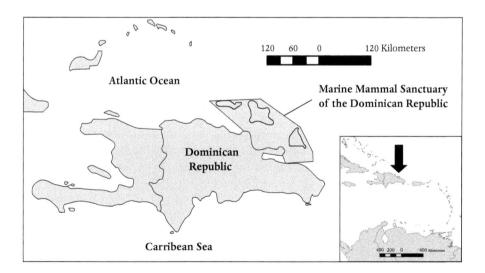

Figure 4.1 Map of the Dominican Republic and the Marine Mammal Sanctuary of the Dominican Republic.

of the beach, with the most desirable beaches going to the tourists. Additionally, many of these resorts and the amenities attached to them, such as stores and golf courses, are completely off-limits to Dominicans, further emphasizing the divide between locals and non-locals.

Although all these resorts employ Dominicans, the companies themselves are foreign owned, and the majority of profits do not stay in the Dominican Republic. This scenario allows for the bulk of revenue to be siphoned toward foreign entities and away from local hotels and restaurants. It creates natural barriers between locals and non-locals, thus unintentionally creating greater social conflict between the groups. For example, many tourists never venture into town without a tour guide. This creates the local perception that tourists find the Dominicans unsafe, while the Dominicans view the tourists as privileged and rich. Many Dominicans simultaneously sell souvenirs to tourists to earn a livelihood while feeling tourists view them as second-class citizens. This distrust is further compounded as tourists are swamped by locals trying to sell trinkets, often at a higher price than local stores. The implications of these interactions will be examined in greater detail below.

This dichotomy of local and non-local participation in the tourism industry is also seen in the whale-watching industry. Tourists often stay in foreign-owned hotels or arrive by cruise ship, both of which provide all-inclusive and/or prepaid packages. Many of these tourists only venture into town for prepackaged whale-watching tours. Those that do arrive in town without a prepaid tour are often inundated by hawkers selling whale-watching tickets on the street. Unlike foreign-owned hotels and cruise ships, many whale-watching businesses are locally owned. Revenue from whale-watching goes directly back into the local economy. Many of these businesses, however, rely on foreign tour operators, who play a large role in setting prices and bringing in clients; this will be discussed later in this chapter.

4.3 The Marine Mammal Sanctuary of the Dominican Republic

Today, the Dominican Republic is the most important whale-watching destination in the Caribbean (Hoyt, 1999). People travel from around the world to view humpback whales (Draheim et al., 2010; Gleason, 2010) and visit pristine beaches (Draheim and Parsons, 2008). In the Dominican Republic, whale watchers view a portion of the North Atlantic humpback whale population in its natural breeding ground. Whales can be found in Silver Bank, Navidad Bank, and Samaná Bay (Clapham et al., 1992; Palsboll et al., 1997; Clapham and Mead, 1999).

Navidad Bank and Silver Bank are located approximately 110 km off of the north coast of the Dominican Republic (Figure 4.1). Silver Bank was originally protected by the Dominican government in 1986 as the Silver Bank Humpback Whale Sanctuary. Within this sanctuary, killing, capturing, or harassing of humpbacks was not permitted. Furthermore, a code of conduct was implemented for whale-watching and some fishing activities (Whaley et al., 2008). In 1996 the sanctuary was expanded to include Navidad Bank, Samaná Bay, and the waters connecting these areas and was renamed the Marine Mammal Sanctuary of the Dominican Republic (MMSRD; Hoyt, 1999). Humpback whales were now protected within the sanctuary and in all surrounding Dominican waters. The expanded sanctuary covered approximately

25,000 km²; but in 2004, for political reasons, Sectorial Law 202–04 was established, which limited the protection of humpback whales to within the sanctuary (Whaley et al., 2008). Unfortunately, this means humpbacks outside the northern waters of the Dominican Republic may have a higher chance of entanglement, boat strikes, and harassment.

In 1985 whale-watching was established at Santa Barbara de Samaná (Samaná for brevity), in the waters of Samaná Bay, Dominican Republic. Whale-watching was available between the months of January and March as a day trip, which offered an alternative to the live-aboard whale-watching trips to the more distant sites of Silver Bank and Navidad Bank. Whale-watching in the area was founded by Kim Beddall, an expatriate from Canada who still lives and operates a whale-watching company in Samaná. Since 1985 whale-watching has steadily grown. In 1985 approximately 165 passengers were taken whale-watching by one whale-watching company (Centro para la Conservación y Ecodesarrollo de la Bahía de Samaná y su Entorno, 1999). By 1998, 39 whale-watching boats took over 21,000 tourists whale-watching (Centro para la Conservación y Ecodesarrollo de la Bahía de Samaná y su Entorno, 1999). Today, over 40,000 people visit the MMSRD during whale-watching season (Agroforsa, 2012) via 33 whale-watching companies and 46 permitted vessels (O'Connor et al., 2009).

In 1992, over concern that whale-watching trips were harassing whales, two nonprofit organizations drafted whale-watching guidelines that were adopted by the boat-owners association two years later. Compliance with these regulations was low, and many felt this was potentially due to the lack of an enforcing body. However, there was also a lack of local involvement prior to the actual adoption of regulations. Only a select few locals were a part of the process. Without such involvement in the decision-making process, stakeholders have little ownership over rules and guidelines, and it is unlikely that guidelines will be upheld (Reed et al., 2008). In the Dominican Republic, it wasn't until a series of tourist complaints, boat accidents, and the departure of a major foreign tour operator from the region that a more formal system evolved (Leon, 2004). In 1998 a co-management system was formed between the Dominican Ministry of Environment, the Dominican Ministry of Tourism, the whale-watching boat-owners association ASDUBAHISA, the Dominican navy, and the Centro para la Conservación y Ecodesarrollo de la Bahía de Samaná y su Entorno (CEBSE), a local marine-conservation nonprofit. This system created a permit system, a monitoring system, a surveillance system, and a fundraising scheme to maintain it (Leon, 2004; Whaley et al., 2008).

Unlike the case in whale feeding grounds located in U.S. waters, all whale-watching in the Dominican Republic must be done on a permitted vessel. For many years, 41 permits were issued, with preference toward vessels previously permitted; however, in recent years that number has fluctuated (Leon, 2004; Ministerio de Turismo de República Dominicana, 2009). Furthermore, with the construction of a new marina, a small number of day passes allowing private vessels to whale watch have been created (personal communication). Under the co-management system, all permitted vessels must meet minimum requirements such as having life jackets and a VHF radio and meet a required minimum vessel size. Dominican naval personnel and representatives from Ministry of Environment inspect vessels to make sure the vessels meet permitting requirements. Once permitted, a vessel is issued a flag which is flown during whale-watching to prove that the boat is legal. The sale of these permits helps to fund the co-management system.

The co-management surveillance system requires permitted vessels to allow an observer on their boat to ensure compliance with whale-watching regulations, and an administrator oversees the vessel observers. Historically, the administrator and observers were politically appointed (no experience necessary), but today the administrator hires qualified observers. While on board, these observers note violations of the whale-watching code of conduct; in addition, they help coordinate daily operations (Leon, 2004). Although there has been improvement in this area, the previous lack of trained observers has created social conflict between the whale-watching captains and current observers.

To ensure whale-watching is not harming the local whale population, CEBSE oversees the monitoring of whales in Samaná Bay. Volunteers board whale-watching vessels and note basic whale behaviors and the location of the whales, as well as data concerning the weather and the length of observation. If possible, CEBSE observers take pictures of flukes (tails) and dorsal fins for the identification of humpback whales. These data are then entered into a database maintained by CEBSE.

Conflict often occurs when stakeholders have different viewpoints, beliefs, and goals regarding a managed resource (Madden, 2004). These conflicts tend to be emotionally charged and have a history of past disputes that include differences in socioeconomic standings and cultural norms (Madden, 2004; Wieczorek Hudenko, 2012). When those in charge of a protected area or resource do not address human and wildlife conflict, the conflict will only increase from a human–wildlife conflict to a human–human conflict (Madden, 2004). Such is the case in the Dominican Republic.

4.4 The dispute over whale-watching

In Los Haitises National Park, tourists can explore caves with pre-Columbian artwork, some of which depicts humpback whales. Humpbacks have been present in Dominican waters for hundreds of years but were never viewed as a valuable meat commodity and therefore were not commercially hunted. Although there are documented cases of opportunistic hunts, humpbacks were primarily ignored until their economic importance through whale-watching was recognized (Leon, 2004). In fact, when Leon (2004) conducted her research in Samaná, she was repeatedly told a story regarding a humpback breaching (jumping from the water) in front of the town in the 1960s. People were so unfamiliar with the creature they thought it was a sign that the end of the world was coming. At the time, even the people most familiar with humpbacks, artisanal fisherman, avoided contact with them out of fear. This is in stark contrast to many other situations, where historical struggles over fisheries have prompted high levels of conflict over marine mammals. For example, the same population of whales prompts conflict between whale conservationists and fishermen in the whale feeding grounds off the northeast coast of the United States, which has an extensive history of commercial fishing. In the Dominican Republic, on the other hand, controversy over whales didn't truly begin until they were seen as an economic commodity by foreign investors. Kim Beddall, who originally arrived in the Dominican Republic while working for a company trying to establish a local scuba-diving business, was the first to view humpbacks as providing potential business.

Today Samaná is a small town which primarily relies on tourism to ensure economic prosperity. A majority of tourists come to Samaná for the sole purpose of watching whales (Gleason, 2010). This reliance on tourism, specifically whale-watching, has the potential to create disputes within the community. Although there is a well-organized and well-thought-out co-management structure on paper, there is still conflict and misunderstanding both within and outside the whale-watching co-management system. The primary dispute in Samaná is over how to manage whale-watching, as different stakeholders see the management of whales differently. Some believe whales should be managed as an economic commodity, others have more conservation-minded beliefs, and still others fall somewhere in the middle. This leads to many current concerns, including illegal whale-watching (discussed in Section 4.5).

Many societal roles within the Dominican Republic contribute to the current conflict over whale-watching. These preconceived societal roles and personnel experiences play a role in Dominicans' attitudes and opinions toward whale-watching (Dickman, 2010). For instance, Dominicans who have greater economic stability may be more prone to adopt conservation viewpoints, while those who struggle financially may only view whales as a means to feed their families. Of course, the opposite might also be true. Those that are economically dependent on whale-watching may be more likely to conserve whales so that the population prospers in Samaná Bay for years to come, while those that are more economically stable, such as boat owners, may tend to adopt the "tragedy of the commons" approach described by Hardin (1968), utilizing whales until the whales' demise and reaping benefits as quickly as possible. Social and natural scientists agree it is important to manage a natural resource for both economic prosperity and conservation (Brenner et al., 2010). This ideal, however, is difficult in reality to carry out and often creates disputes within the community. This is especially true when dealing with the tourism trade. Sustainable tourism is hard to create and maintain because the tourism resource, in this case, whales, is a common-pool resource (Briassoulis, 2002). Thus, it is difficult to prevent stakeholders from utilizing the resource as they see fit, let alone managing outside interest groups such as cruise ships (discussed in Section 4.7).

As with most conflicts, there is a history of underlying and identity-level factors beneath the surface. Without recognizing conflict at these deeper levels, it is impossible to resolve present-day concerns, as past problems continue to arise and give meaning to the current dilemma (Chapter 1; Madden and McQuinn, 2014). The following sections will explore the deeper levels of conflict surrounding the primary dispute of how whale-watching and whales should be managed in the MMSRD. These concerns will be examined in three categories: conflicts on the water, conflicts off the water, and new conflicts.

4.5 Conflicts on the water

4.5.1 Underlying conflict

Illegal whale-watching boats have always been a concern within the MMSRD. The co-management system creates a working permit system, but many people are excluded. Although most locals are aware that you must have a permit to watch whales, they may not have enough money to purchase a permit and may therefore make the decision to work outside the law. These operators tend to own one or two small boats because maintaining a large whale-watching vessel (Figure 4.2), that complies with regulations is expensive. The permit system creates an exclusive organization for those that can afford the permits and boats and thereby creates an unequal power structure in the community. This power structure goes unchallenged, as those who can afford a permit have a voice in the decision-making process within the co-management system, while those outside it do not. (And this power asymmetry further leads to identity conflict, discussed in Section 4.6.2.) A previous evaluation of the co-management system showed that, over the years, tensions with small-boat owners increased as the co-management group slowly increased the required size of a permitted whale-watching vessel (Leon, 2004). Whether small boats are illegal or permitted, however, they are the easiest way for most locals, who tend to not be wealthy, to access the whale-watching industry. When small vessels are excluded, an entire group of stakeholders is removed from the legal whale-watching industry. This creates more tension regarding the dispute of how whales and whale-watching should be managed.

A primary management technique employed by the co-management system is the use of observers on whale-watching vessels to determine if guidelines are being followed. Although the Dominican navy is the local enforcement agency for the guidelines, they are rarely on the water to view infractions. They rely almost entirely on the observers' reports. In the recent

Figure 4.2 Typical whale-watching vessels awaiting passengers on a dock in Samaná Bay. Courtesy of Christine Gleason.

past, observers have been political appointees. These observers were primarily not from Samaná, were unfamiliar with the management system, and instead of observing concerns on the water, spent much of their time sleeping on the vessels. As observers were not reporting vessels that were breaking the guidelines, the captains were doing more and more of the observers' job. Over the past several years, observers have gained more control, primarily due to consistent administrative leadership, but there are still times their warnings go unheeded. This underlying conflict has created an identity-level conflict in which boat captains assume that all observers will fail to do their jobs.

4.5.2 Identity-level conflict

Underlying conflict on the water has led to identity-level conflict between small and large boats and between boat captains and observers. The lack of voice and unequal power between small- and large-boat owners has created a deep lack of trust between the two groups. In the past, small boats have created bad press for the whale-watching industry. This is due to small-boat incidents involving a lack of safety precautions, as well as to two accidents that caused injuries and left tourists adrift for hours (Leon, 2004). These incidences led to a large foreign tour company refusing to allow their clientele to whale watch in Samaná. This reduced business for the entire industry. Large-boat captains now identify all small-boat captains and owners as problematic and expect smaller vessels to act unsafely. Large-boat owners then lobby for small-boat owners and operators to not receive permits. The cycle continues to perpetuate itself; small boats are excluded, do not receive permits, and, as illegal boats, act unsafely, which increases the perceived identity-level conflict that all small boats are unsafe. These incidences on the water are in addition to the lack of compliance with boat permitting and in many cases whale-watching regulations.

In order to manage the lack of compliance and illegal boats, observers are used. However, boat captains make assumptions about new observers based on their past experiences (Dickman, 2008). Even though there has been considerable improvement in this area, they still identify current observers with those that did not do their jobs. This identity-level conflict further impedes the co-management systems' ability to manage whale-watching within the sanctuary. Overall, the idea of observers on the vessels should allow for better guideline enforcement, but the lack of trust between observers and boat captains may negate any benefits.

4.6 Conflict off the water

4.6.1 Underlying conflict

There are five groups within the management system: the Ministry of Environment, the Ministry of Tourism, the whale-watching boat-owners association, the Dominican navy, and CEBSE. Each group has varying goals for whale management, but there are also subdivisions along non-organizational lines. Leon (2004) describes a situation that occurred regarding the annual revisions to whale-watching guidelines. The 2004 revisions were faxed from the Ministry of Environment in Santo Domingo to a select group of boat owners. There had been no previous stakeholder input, and the revisions were faxed only a few days prior to the start of the whale-watching season. Although comments by fax were welcomed, there was no time provided to review these comments or make changes prior to the updated regulations going into effect for the season. This meant that only those who had greater resources (such as fully equipped offices with fax machines) had the opportunity to view the revisions. Moreover, it added to the increasing discontent from other groups within the co-management system.

Although overall most boat owners were happy, many expressed concerns that the ministry was making decisions without consulting all stakeholders. As the conflict between pro-conservation and pro-business managers increased, the ministry made more decisions aimed at increasing conservation protections but with no stakeholder input. This left other co-managers, such as the boat-owners association, feeling less empowered. Because they had no clear voice or formal power in the revision process, they became less motivated to follow the guidelines or be part of creating new guidelines in the future (Madden and McQuinn, 2014). Currently, an effort to work more collaboratively with the Ministry of Environment and the sanctuary administrator has improved relations. Being able to unify these varying viewpoints, where both sides benefit, could go a long way in mitigating the conflict within the co-management system (Treves et al., 2006).

4.6.2 Identity-level conflict

Many boat owners live in Samaná full time but may have been born elsewhere. They tend to have higher levels of education and greater economic resources than the general public of Samaná. Their social standing allows them easier access to co-management representatives in both Samaná and Santo Domingo. Boat captains, crews, and smaller boat owners, however, tend to be members of the working class and have little input in the co-management system. This divide creates a lack of information flow between the two social groups. This is unfortunate, as boat captains and crews have the most on-the-water experience and could have valuable input. In addition, many small-boat owners are in need of education regarding guidelines but cannot access training as they are (primarily) not a part of the co-management group. Neither group is fully represented in the co-management system, so the system doesn't benefit from the experienced crew and captains or have the buy-in necessary for the small-boat operators to abide by the guidelines. The lack of full representation of these working classes leads to further conflict within the co-management system. Compounding this is the fact that these groups do not overlap in everyday life. For example, large-boat owners and ministry representatives (those with greater resource access) can be seen socializing together at one location, whereas boat captains, crews, and small-boat owners are seen in another. This economic and social divide is based on a history of identity-level conflict from colonization times that persists today.

The Dominican Republic has a long history of colonization and slavery, and the majority of the Dominican's population is mulatto or of mixed-race ancestry (Fuller, 1999). This has set precedence within the country regarding social class and skin tone (please note these are generalizations and do not apply to every situation). Lighter-skinned Dominicans tend to be economically more established and have greater access to higher education. This allows these individuals to gain entry to more influential jobs, including those in the government that can help to increase their wealth. Darker-skinned Dominicans tend to be viewed as working class. They may have been unable to finish school, let alone have access to higher education and therefore higher-paying jobs. Dominicans, in general, view those with lighter skin as better educated and more economically stable.

This type of social and economic segregation is seen in the conflict between the small-boat owners, large-boat owners, and the ministry of the environment. People with greater influence, such as large-boat owners and ministry officials, are making decisions impacting those who do not have a clear voice in the process. Over time, the minimum size of a permitted whale-watching boat has increased, excluding small-boat owners and operators. Although Leon (2004) notes in her evaluation that greater stakeholder collaboration was needed, especially in decision-making processes, this is difficult to create with these social divides. It requires an approach that transforms the identity-level conflicts.

These efforts can be made even more difficult in the Dominican Republic due to the country's long history of colonization, occupation, and slavery, which has created an inherent mistrust of all outsiders (see Comment). This manifests as suspicious attitudes toward those from outside the country, not from Samaná, or of different social classes. The Dominican's history has fostered a sense of national pride, which in turn creates an inclination for Dominicans to want to accomplish all goals with little (if any) outside help. This can be seen on both national and international levels. Internationally, for example, if foreign researchers want to work with a community, it can be very difficult to obtain the necessary permits. In order to submit a permit for potential research, you must turn in an application to the Ministry of Environment in person, which raises the financial and logistical costs of working there. This can prevent important research, which could benefit the whale-watching community, from occurring. Research, such as assessing the impacts of whale-watching on whale populations, could be invaluable to preserving a healthy whale population and thus, in turn, would continue to support whale-watching and economic prosperity within the town. Therefore, difficulties in obtaining research permits could harm the co-management system.

Comment

For another example of how a history of colonialism and oppression can shape modern-day conflict, see Julie-Beth McCarthy's chapter on religion and human–wildlife conflict (Chapter 9). In that case, the Bajau, an indigenous group that sought to avoid conversion to Islam in the fifteenth century, became sea nomads, a move that has put them at odds with present-day marine managers.

—The Editors

Nationally this independent spirit can be seen (as is the case in Samaná) in social divisions. In many cases, people from areas such as the capital are seen negatively until they have proven themselves otherwise. Within the co-management system, representatives tend to be well-educated individuals who may or may not be from Samaná. For example, CEBSE is the "local" nonprofit in Samaná, but the person who oversees CEBSE lives in Santo Domingo. Only during whale-watching season is there a local representative. Even the administrator for the local sanctuary, who retired in 2014, spent several years dwelling in the capital and commuting for the whale-watching season. It is only in the most recent years that this has changed. This makes certain representatives less accessible to stakeholders, creating a lack of communication and an unequal distribution of power within the system. Furthermore, it creates distrust between the stakeholders. These outsiders are viewed as rule-makers trying to regulate resources within Samaná, and even if what they are doing has positive impacts, it may be viewed negatively, given past prejudices. For example, when an upper-class, light-skinned Dominican from Santo Domingo was given the job as the ministry administrator, it took several years for him to build any level of trust with all social classes. This was further compounded by the fact he was not a local. As this trust has grown, the administrator attempts to work between social and economic divisions, but much more needs to be addressed to resolve the conflict regarding whale-watching management in Samaná Bay.

4.7 New concerns on the water

Recently, new entities within the whale-watching industry have created additional areas of concern and conflict that were not present when the original management scheme was created. The presence of cruise ships in Samaná Bay has greatly increased (Figure 4.3). Although many welcome this influx of economic growth, there is great concern that these ships will harm the whales through either direct strikes or pollution. This issue divides the community both within and outside the co-management framework. This adds to the dispute as to how to manage whale-watching and whale conservation in Samaná Bay.

4.7.1 New disputes

Cruise ships contract with specific whale-watching companies within the boat-owners association. Many of these contracted companies must follow cruise-ship guidelines as far as start and end times, as well as the length of trip. Cruise ships' guided tours tend to be short in order to allow tourists more time for fun in the sun on a local island. In order to keep the tourists and cruise ships happy, boat captains may break guidelines by approaching whales too quickly or by getting too close. This occurs because finding whales on a shorter trip can be difficult. If a boat captain hears of a whale's location over the radio, he may break speed regulations to find it and then not wait his turn to view the whale if other boats are already present. This behavior is allowed to continue, as the cruise ships bring business to the large whale-watching companies, even though the behavior might threaten the very commodity they have come to watch.

Figure 4.3 Cruise ships off a small island (Cayo Levantado) near Samaná, at the far end of the Marine Mammal Sanctuary of the Dominican Republic. Courtesy of Christine Gleason.

This creates concerns between those who make a living from the cruise ships and those who refuse to cater to the cruise-ship trip format. In fact, these differently-structured trips seem to epitomize the fact that not everyone views whale-watching and/or the tourism industry in the same way (Bimonte, 2008). Those who work with cruise ships refuse to change their trip formats or insist that their boat captains follow guidelines. Moreover, outside the whale-watching industry, many locals rely on cruise ships to bring in tourists to earn money by selling souvenirs or offering tours to other tourist locations. While the whale-watching trips contracted by cruise ships, as well as those locals who sell souvenirs, welcome cruise ships, those concerned about whale conservation and the sustainable continuation of the whale-watching industry do not.

4.7.2 Underlying conflict

These new disputes have led to underlying conflict on the water. On one occasion, the author personally observed boat operators who relied on cruise ships prevent a boat that did not from docking and picking up a small number of cruise-ship passengers. These passengers had decided not to utilize the prepackaged arrangement via the cruise ship and to book with an organization that had an interpreter aboard. The boat that was not allowed to pick up passengers was a conservation-minded company that viewed cruise ships as bad for the whale-watching business. Unfortunately, this was not an isolated incident. These negative interactions accumulate over time, leading to a greater atmosphere of distrust. Ultimately, this leads to identity-level conflict.

4.7.3 Identity-level conflict

The ongoing and past action of this new actor—cruise ships—has led to an identity-level conflict. Those that view cruise ships as a potential threat to whale conservation, and therefore whale-watching, make assumptions about cruise ships that color interactions in a negative way. This type of action, although only practiced by some of the cruise-ship whale-watching boats, has reinforced preconceived notions that cruise ships are bad.

But this is only one of many new identity-level conflicts in Samaná. As whale-watching has grown, so has the general tourism trade within the town. Shops have popped up along the main walkway, and vendors set up tents when cruise ships enter the town. Even children can be seen selling local food and music. These sellers, however, do not speak fluent English and often aggressively approach tourists. This leads to tourists feeling like they are being mobbed and/or swindled. Tourists often refuse to buy items directly from these locals and instead purchase them in a more traditional store, owned by the foreign hotel, in part because they believe that all locals are untrustworthy. Likewise, locals take this as further evidence that tourists view themselves as higher class. Each group now identifies with these preconceived perceptions, creating identity-level conflict. Similar situations can be seen in the whale-watching industry. Local, often illegal, boat crews approach tourists on the street and try to sell tourists a whale-watching trip, which is generally in a small boat. This leads to illegal boats on the water and thus to conflict between legal and non-legal vessels. This is why it is essential that all stakeholders be integrated into the co-management process to ultimately transform all levels of conflict that are negatively impacting effective whale-watching management in Samaná.

4.8 Conclusions

Hardin (1968) argued that the only way to truly avoid the tragedy of the commons was to allow for private control over the resource. Today, others argue that it is only by considering the viewpoints of all stakeholders that true collaborative management can occur and the tragedy can be averted (Ostrom, 1990). This is especially important when you have a group of stakeholders that are not homogenous in their viewpoints and backgrounds and who may all envision the resource and its management differently (Adams et al., 2003; Carlsson and Berkes, 2006). In these situations, conflicts can, and do, occur. Despite these complexities, however, the co-management of common-pool resources, such as a population of whales for sustainable whale-watching, is possible (Dietz et al., 2003). No management system, however, is perfect, and constant monitoring and adaptation in response to the social conflict dynamics and the needs within the system is essential if management is to succeed (Ostrom et al., 1999; Adams et al., 2003).

In 2004, an assessment of the co-management system was conducted. Although it found that people were primarily in favor of the system, several participants noted the need for better collaboration. Furthermore, this evaluation found that if a dispute arose within the system, there was no formal process to resolve it. No system is static. Opinions, politics, and scenarios change and conflicts arise. Without a formal and ongoing conflict transformation strategy, tensions, as seen throughout this chapter, can arise within the system, and a lack of compliance with co-management guidelines can occur. Although CEBSE has stepped in as a moderator in the past, it was a temporary solution only.

With social-class divisions and inherent mistrusts, it is difficult to find a neutral party moderator to resolve these conflicts. Even if this moderator has no stake in the outcome of the process, if they are not local and/or fall within the correct social-class divisions, they are not seen as a legitimate neutral entity. This adds another layer of mistrust that a neutral moderator must overcome to earn the trust of all parties. Furthermore, the Dominican Republic is a developing country and does not necessarily have the economic resources to re-evaluate the system. Therefore, there is a need, if not a want, for outside help, if only to build local capacity to effectively reconcile the social conflicts and design and facilitate a process to ensure a more sustainable co-management system. It is the conclusion of this author that without a new and more comprehensive evaluation of the co-management system and the development of a conflict transformation process, conflict will continue to increase and the management system in place will continue to falter. This will ultimately have a negative effect on whale-watching and whale conservation in Samaná Bay. Any new evaluation within the co-management system needs to ensure a genuinely inclusive and empowering process for all groups. Stakeholder engagement efforts need to focus on all levels of the conflicts and not just the dispute. The deeper-rooted social conflicts can prevent sustainability of the industry and genuine commitment to co-management efforts. By contrast, transforming these conflicts often results in greater creativity and commitment to management efforts, as can allowing stakeholders, including those not currently a part of the co-management system, an opportunity to voice their concerns (Chapter 1; Madden and McQuinn, 2014). Without a level of trust between managers, a lack of investment in and adherence to guidelines will occur (Madden and McQuinn, 2014).

Within the Dominican Republic, the co-management system helped to organize stakeholders involved in whale-watching and provide whale protection. The last evaluation of this system, however, occurred almost ten years ago. Since that time many new concerns such as those generated by the presence of cruise ships have arisen, and many underlying and identity conflicts still exist. Although the co-management guidelines are reviewed yearly, they are reviewed within the co-management system, leaving little ability for outside stakeholders to comment. Many other entities have knowledge or concerns that should also be considered when guidelines are revised. For example, small-boat owners not already part of the co-management system, small-boat captains and crews, those who contract cruise ships to enter the port, those who sell souvenirs, or those who own local hotels should all have a voice in the process. All of these factions would lose their primary business if whale management were not done properly. Therefore, it is important to adopt a new component to the co-management system that would allow for all stakeholder groups to express concerns. By working to transform the co-management system, deeper underlying and identify-level conflicts can be addressed, and more sustainable co-management of whales for both conservation and economic prosperity can occur.

Lessons learned

- Be aware of policies and processes that create unequal power structures, as these can increase conflict over time.
- Tourism can be a driver of conflict, as people with competing economic interests clash.
- Common-pool resources require special consideration. Conflict over these resources can most effectively be resolved or transformed through addressing the social aspects of the conflict.

—The Editors

References

Adams, W. M. Brockington, D., Dyson, J., and Vira, B. (2003). Managing tragedies: understanding conflict over common pool resources. *Science*, **302**(5652):1915–16.

Agroforsa. (2012). *Estudio del conocimiento, percepción, actitud y aportes económicos de seis áreas protegidas seleccionadas de Bahía de Samaná y su entorno*. Report to the Nature Conservancy and the United States Agency for International Development. Ensanche Naco Santo Domingo, República Dominicana: Agroforsa.

ASONAHORES. (2003). *Estadisticas seleccionadas del sector turism ano 2002*. Santo Domingo, República Dominicana: ASONAHORES.

Berkes, F., George, P., and Preston, R. (1991). Co-management: the evolution of the theory and practice of joint administration of living resources. *Alternatives*, **18**(2):12–18.

Bimonte, S. (2008). The "tragedy of tourism resources" as the outcome of a strategic game: a new analytical framework. *Ecological Economics*, **67**(3):457–64.

Brenner, J., Jiménez, J. A., Sardá, R., and Garola, A. (2010). An assessment of the non-market value of the ecosystem services provided by the Catalan coastal zone, Spain. *Ocean and Coastal Management*, **53**(1): 27–38.

Briassoulis, H. (2002). Sustainable tourism and the question of the commons. *Annals of Tourism Research*, **29**(4):1065–85.

Carlsson, L. and Berkes, F. (2006). Co-management: concepts and methodological implications. *Journal of Environmental Management*, **75**(1):65–76.

Centro para la Conservación y Ecodesarrollo de la Bahía de Samaná y su Entorno (CEBSE). (1999). *Diagnóstico del sector turismo de la provincia de Samaná*. Samaná, República Dominicana: CEBSE.

Clapham, P. J. and Mead, J. G. (1999). *Megaptera novaeangliae*. *Mammalian Species*, **604**:1–9.

Clapham, P. J., Palsbell, P. J., Mattila, D. K., and Vasquez, O. (1992). Composition and dynamics of humpback whale competitive groups in the West Indies. *Behavior*, **122**(3):182–94.

Dickman, A. J. (2008). *Key Determinants of Conflict between People and Wildlife, Particularly Large Carnivores, around Rugha National Park, Tanzania*. London: University College London.

Dickman, A. J. (2010). Complexities of conflict: the importance of considering social factors for effectively resolving human–wildlife conflict. *Animal Conservation*, **13**(5):459–66.

Dietz, T., Ostrom, E., and Stern, P. C. (2003). The struggle to govern the commons. *Science*, **302** (5652):1907–12.

Draheim, M., Bonnelly, I., Bloom, T., Rose, N., and Parsons, E. C. M. (2010). Tourist attitudes towards marine mammal tourism: an example from the Dominican Republic. *Tourism in the Marine Environment*, **6**(4):175–83.

Draheim, M. and Parsons, E. C. M. (2008). Hispaniola. In Luck, M. (ed.), *Encyclopedia of Tourism and Recreation in Marine Environments*. Cambridge, MA: CABI, pp. 214–15.

Fuller, A. (1999). *Tourism Development in the Dominican Republic: Growth, Costs, Benefits and Choices*. <kiskeya-alternative.org/publica/afuller/rd-tourism.html>, accessed June 16, 2013.

Gleason, C. M. (2010). *The Conservation Awareness and Attitudes of Whale-Watching Tourists in Samaná, Dominican Republic*. Keene, NH: Antioch University New England.

Hardin, G. (1968). The tragedy of the commons. *Science*, **162**(3859):1243–8.

Hoyt, E. (1999). *The Potential of Whale Watching in the Caribbean: 1999+*. Bath: Whale and Dolphin Conservation Society.

Lawrence, T. B., Philips, N., and Hardy, C. (1999). Watching whale-watching: exploring the discursive foundation of collaborative relationships. *Journal of Behavioral Science*, **35**(4):479–502.

Leon, Y. M. (2004). *community Impacts of Coastal Tourism in the Dominican Republic*. <http://digitalcommons.uri.edu/dissertations/AAI3147809>, accessed March 11, 2015.

Madden, F. (2004). Creating coexistence between humans and wildlife: global perspectives on local efforts to address human-wildlife conflict. *Human Dimensions of Wildlife*, **9**(4):247–57.

Madden, F. and McQuinn, B. (2014). Conservation's blind spot: the case for conflict transformation in wildlife conservation. *Biological Conservation*, **178**:97–106.

Ministerio de Turismo de República Dominicana. (2009). *Permisos solicitados para en la temporada de observacion de ballenas 2009, Bahia de Samana, Santo Domingo, Dominican Republic*. Santo Domingo, Dominican Republic: Ministerio de Turismo de República Dominicana.

Moore, S. A. and Rodger, K. (2010). Wildlife tourism as a common pooled resource issue: enabling conditions for sustainability governance. *Journal of Sustainable Tourism*, **18**(7):831–44.

Neves-Graca, K. (2004). Revisiting the tragedy of the commons: ecological dilemmas of whale-watching in the Azores. *Human Organization*, **63**(3):289–300.

O'Connor, S., Campbell, R., Cortez, H., and Knowles, T. (2009) *Whale-Watching Worldwide: Tourism Numbers, Expenditures and Expanding Economic Benefits. A Special Report from the International Fund for Animal Welfare*. <http://www.ecolarge.com/wp-content/uploads/2010/06/WWW09.pdf >, accessed March 11, 2015.

Ostrom, E. (1990). *Governing the Commons: The Evolution of Institutions for Collective Action*. Cambridge: Cambridge University Press.

Ostrom, E., Burger, J., Field, C. B., Norgaard, R. B., and Policansky, D. (1999). Revisiting the commons: local lessons, global challenges. *Science*, **284**(5412):278–82.

Palsboll, P. J., Allen, J., Berube, M., et al. (1997). Genetic tagging of humpback whales. *Nature*, **388**(6644):767–9.

Reed, M. S., Dougill, A. J., and Baker, T. (2008) Participatory indicator development: what can ecologists and local communities learn from each other? *Ecological Applications*, **18**(5): 1253–69.

Tejada, V. M. (1996). Los nuevos conquistadores. La industria hotelera no es tan criolla. *Rumbo*, **612**:42–8.

Treves, A., Wallace, R. B., Naughton-Treves, L., and Morales, B. (2006). Co-managing human-wildlife conflicts: a review. *Human Dimensions of Wildlife*, **11**(6):383–96.

Weinstein, M. P. and Reed, D. J. (2005). Sustainable coastal development: the dual mandate and recommendation for "Commerce Managed Areas." *Restoration Ecology*, **13**(1):174–82.

Whaley, A. R., Wright, A. J., Di Calventi, I. B., and Parson, E. C. M. (2008). Humpback whale sightings in southern waters of the Dominican Republic lead to proactive conservation measures. *Marine Biodiversity Records*, 1:e70.

Wieczorek Hudenko, H. (2012). Exploring the influence of emotion on human decision making in human–wildlife conflict. *Human Dimensions of Wildlife*, **17**(1):16–28.

World Bank. (2012). *Dominican Republic*. <http://data.worldbank.org/country/dominican-republic#cp_wdi>, accessed June 16, 2013.

5

Levels of Marine Human–Wildlife Conflict

A Whaling Case Study

E. C. M. Parsons

5.1 Introduction

Conflict transformation theory, as adapted to human–wildlife conflict by Madden and McQuinn (2014; also see Chapter 1), identifies three levels of conflict. These three levels of conflict are evident in the controversy over Japanese "scientific" whaling at the International Whaling Commission (IWC).

5.1.1 The IWC

In response to declining catches and sightings of whales, in 1931 several nations conducting whaling operations joined together to enact the *Convention for the Regulation of Whaling* in an attempt to manage and ensure the sustainability of the whaling industry. This convention was modified in 1937 and renamed the *International Agreement for the Regulation of Whaling*, and in 1946 it was modified again; the resulting treaty was called the *International Convention for the Regulation of Whaling* (hereafter referred to as "the Convention"; Parsons et al., 2012) and established the IWC as the international body responsible for the management of whale stocks.

In the preamble of the Convention, the need to conserve whale populations for future generations was clearly established:

> Recognizing the interest of the nations of the world in safeguarding for future generations the great natural resources represented by the whale stocks;

> Considering that the history of whaling has seen over-fishing of one area after another and of one species of whale after another to such a degree that it is essential to protect all species of whales from further over-fishing.

Human-Wildlife Conflict: Complexity in the Marine Environment. Edited by Megan M. Draheim, Francine Madden, Julie-Beth McCarthy, and E. C. M. Parsons © Oxford University Press 2015. Published 2015 by Oxford University Press.

The Convention preamble highlights the need to "provide for the proper conservation of whale stocks" but conversely also states that the treaty exists to "make possible the orderly development of the whaling industry"; unsurprisingly, these somewhat contradictory aims have led to much polarization within the IWC, as different factions interpret the text of the treaty in different ways (i.e., the IWC exists to conserve cetacean populations vs. it exists to support, maintain, and ensure commercial whaling; Parsons et al., 2012). Notably, the IWC was one of the first international treaties intended to be guided by science, and the IWC has a scientific committee of 200–400 scientists tasked with advising the commission. Currently, the IWC has 89 member nations, of which slightly more than half arguably have a "pro-conservation" stance as opposed to a "pro-whaling" stance.

In 1982, the IWC voted to instigate a moratorium on commercial whaling due to concerns about the global depletion of whale stocks; this moratorium came into effect in 1986. However, several countries still continue to hunt and kill whales, including Norway, Iceland (both via a "reservation" so they are not bound by the moratorium and can legally harvest whales (MacLeod, 1994; Burns, 1997), although the legality of the latter's reservation is in doubt, e.g., Environmental Policy and Law, 2002; Wansborough, 2004), and Japan (via its so-called scientific whaling, as explained below).

5.1.2 Scientific whaling

Japan joined the IWC in 1951. Despite having a long history of coastal whaling in some locations, Japan was not considered a major whaling nation until after World War II (Ellis, 1991; Scott, 1999; Parsons et al., 2012) and thus was not one of the original signatories of the Convention. The Japanese government initially did not sign the 1982 whaling moratorium but after political pressure and the threat of sanctions (including a loss of fishing rights in Alaskan waters) agreed to abide by the moratorium in 1988 (Parsons et al., 2012). However, the Japanese government continued to catch whales via a loophole in the Convention that allows member nations to catch whales for scientific research purposes. Article 8 of the Convention states that:

> Notwithstanding anything contained in this Convention, any Contracting Government may grant to any of its nationals a special permit authorizing that national to kill, take, and treat whales for purposes of scientific research subject to such restrictions as to number and subject to such other conditions as the Contracting Government thinks fit, and the killing, taking, and treating of whales in accordance with the provisions of this Article shall be exempt from the operation of this Convention. Each Contracting Government shall report at once to the Commission all such authorizations which it has granted. (Art 8, Sec. 1)

There is no limit to the number or type of whales that may be taken; as noted above the take is according to what governments "see fit," and technically even endangered species could be taken. According to the Convention, the carcasses of animals taken for scientific research purposes should "so far as practicable be processed and the proceeds shall be dealt with in accordance with directions issued by the Government by which the permit was granted" (Art 8, Sec. 2). The original intent of this section of the Convention was to reduce wastage of whale products once scientific data were collected and samples were obtained, but in effect this means that the Japanese government can catch whales with a self-allocated quota and sell the resulting whale meat, once what might be token tissue samples and measurements are taken (Parsons et al., 2012). In other words, what is effectively commercial whaling can be conducted via this loophole, but without the burden of a quota system or regulations.

There is no facility under the Convention under which an application for a scientific research permit to catch whales may be turned down, although there is a requirement under Article 8 of the Convention that the data collected by the before mentioned scientific research be presented to the IWC every year (at the meeting of the IWC Scientific Committee).

The Japanese government's self-allocated permit allows for a take of 850 (±10%) Antarctic minke whales (*Balaenoptera bonaerensis*), 50 fin whales (*Balaenoptera physalus*), and 50 humpback whales *(Megaptera novaeangliae;* Figure 5.1) each year in the waters around Antarctica in a program called "JARPA II," although to date no humpback whales have been

Figure 5.1 A humpback whale in Antarctica. For several years the Japanese government has threatened to add humpback whales to their scientific whaling program. The whales are recovering in Antarctica, but high public concern for the whales, which are the focus of whale-watching industries in many parts of the world (especially anti-whaling nations like Australia) also makes them a negotiating item (Photo by E. C. M. Parsons)

captured. Notwithstanding the issue of scientific whaling, these catches are also controversial because since 1994 the waters of the Sothern Ocean around Antarctica have been designated by the IWC as a whale "sanctuary," within which commercial whaling has been banned. Moreover, fin whales are considered endangered by the IUCN (Reilly, 2008a). In addition, the most recent circumpolar surveys (1991–2004) to estimate minke whale numbers found only 40% of the number of minke whales reported in previous surveys (1985–1991; Reilly, 2008b). Some scientists feel that such an observed decline should mean Antarctic minke whales be considered "endangered" under IUCN red data listing criteria, that is, an observed decrease in numbers of more than 50% within a 10-year period (Parsons et al., 2012). In addition the "JARPN II" program in the North Pacific has a self-allocated quota, or sample size, of 340 northern minke whales (*Balaenoptera acutorostrata*), 50 Bryde's whales (*Balaenoptera edeni*), 100 sei whales (*Balaenoptera borealis*), and 10 sperm whales (*Physeter macrocephalus*).

By the beginning of the 2014 meeting, the IWC had passed at least 19 resolutions criticizing the Japanese government over scientific whaling under permit, repeatedly requesting the cessation of this practice (see Box 5.1 for information about the most recent IWC deliberations).

Box 5.1 The 2014 International Whaling Commission meeting[1]

The International Court of Justice (ICJ) ruled that Japan's JARPA II research program was, effectively, illegal.[2] Although as noted in this chapter, Article VIII of the International Convention on the Regulation of Whaling allows the lethal take of whales for scientific research purposes by "special permit," the ICJ ruled that the Japanese program was in violation of this provision because JARPA II was not bona fide scientific research but was instead de facto commercial whaling.

The Japanese government initially stated that it would abide by the ICJ's decision and discontinue JARPA II but then later announced it would conduct a new research program in the Antarctic. This about turn may have also been influenced by Sea Shepherd publicly claiming to have "defeated" the Japanese government and that they had played a major role in forcing them to end the Antarctic hunt.[3] (Note that Sea Shepherd was not involved in the ICJ court case and can claim no responsibility for the outcome.) As stated in this chapter, for the fiercely proud and nationalistic Japanese politicians to have a small NGO—which they have labeled a "terrorist organization"—beat them would be cause a loss of face and thus be politically untenable.

The ICJ ruling became a point of significant controversy at the International Whaling Commission (IWC) meeting in September 2014. At the meeting, the New Zealand government spearheaded a resolution that was passed by the majority of IWC Commissioners (IWC/65/14 Rev 1),[4] which instructed the IWC Scientific Committee to determine:

[1] A longer version of this textbox article was first published by the author on the marine science blog site *Southern Fried Science* (<http://www.southernfriedscience.com/?p=17800>).

[2] The judgment can be found at <http://www.icj-cij.org/docket/files/148/18136.pdf>. A summary press release on the judgment can be found at <http://www.icj-cij.org/docket/files/148/18162.pdf>.

[3] For example, see <http://www.seashepherd.org/news-and-media/2014/03/31/the-whales-have-won-icj-rules-japans-southern-ocean-whaling-not-for-scientific-research-1569>and<http://www.seashepherd.org/commentary-and-editorials/2014/04/04/a-new-chapter-in-whale-conservation-643>.

[4] Downloadable from <https://archive.iwc.int/pages/view.php?ref=3452>.

continued

(a) whether the design and implementation of the [scientific whaling] programme, including sample sizes, are reasonable in relation to achieving the programme's stated research objectives;

(b) whether the elements of the research that rely on lethally obtained data are likely to lead to improvements in the conservation and management of whales;

(c) whether the objectives of the research could be achieved by non-lethal means or whether there are reasonably equivalent objectives that could be achieved non-lethally;

(d) whether the scale of lethal sampling is reasonable in relation to the programme's stated research objectives, and non-lethal alternatives are not feasible to either replace or reduce the scale of lethal sampling proposed; and

(e) such other matters as the Scientific Committee considers relevant to the programme, having regard to the decision of the International Court of Justice, including the methodology used to select sample sizes, a comparison of the target sample sizes and the actual take, the time frame associated with a programme, the programme's scientific output; and the degree to which a programme coordinates its activities with related research projects.

The resolution also called for the IWC Scientific Committee to revise its review procedure for lethal research proposals. The process in recent years has been made somewhat controversial by the participation of the (scientific whaling) proposal proponents in the review of their own proposals (ignoring the obvious conflict of interest), and their comments receiving equal weight to those of the proposals' critics within the Scientific Committee. It has been argued that this is akin to high-school students being able to grade their own essays, and their grades being given equal weight to those of the teacher when assessing scores. Bringing IWC practices in line with standard academic review procedures was something that was considered to be essential for a treaty that is supposed to be advised by science.

Immediately after the resolution was passed, the Japanese government announced that it will be submitting a new research proposal for Antarctica that will be "in line with international law as well as ICJ language" in 2015.

5.2 The dispute

As described by Madden and McQuinn (2014), the first level of human–wildlife conflict is "the dispute" itself, that is, the current, tangible issue or problem that is in contention. In this case, the main dispute is about the validity and value of the scientific research conducted by special permit, or so-called scientific whaling. Some of the controversial aspects of scientific whaling have been described above, but the main dispute comes down to an argument about science, as stated by the IWC (2013):

> There has been and remains considerable disagreement over the value of this research both within the Scientific Committee and the Commission. Particular disagreement within the Committee has focussed on a number of issues, including: the relevance of the proposed research to management, appropriate sample sizes and applicability of alternate (non-lethal) research methods.

The JARPA and JARPN programs of Japan have been heavily criticized by scientists, including those within IWC's own scientific committee (Clapham et al., 2003; Gales et al., 2005;

Clapham et al., 2007; Corkeron, 2009; Figure 5.2). Many argue that the data gathered by scientific whaling are not required for the management of whale populations, that sampling regimes are inappropriate with excessive numbers of whales being taken, that much of the data can be gathered by non-lethal means and, ultimately, that the quality of the science is poor, with Clapham et al. going so far as to state: "many [IWC Scientific Committee] members have contended that Japan's scientific whaling program is so poor that it would not survive review by any independent funding agency" (Clapham et al., 2003, p. 212). Conversely, Japanese representatives argue that the large numbers of whales taken is required to achieve statistical validity, that the data that need to be gathered (such as diet information from stomach contents) can only be collected via lethal methods, that the program is perfectly legal and in accordance with the Convention, and that the program has produced over a hundred publications, some in international journals, undermining the belief that their science is poor (Hatanaka, 2005).

Morishita (2006) summarized this more tangible aspect of the situation:

> The whaling issue can be described as a scientific dispute over resource management including issues of stock abundance and scientific uncertainties. (p. 802)

However, Morishita (2006) viewed the debate over science as being somewhat spurious, arguing that whale stocks were recovering, that the IWC had a robust science-based quota system (the Revised Management Procedure, or RMP, although scientific whaling is not covered or regulated by this quota system), and that takes via special permit, or scientific whaling, were at levels that populations could sustain. He also argued that part of the dispute was not on scientific grounds, but on economic grounds:

> Another aspect of the whaling controversy is economics. In the past, the economic issue was the survival of whaling industry. However, now the whaling issues are used by anti-whaling organizations as a tool to raise funds for the organization. (Morishita, 2006, p. 802)

The national governments and NGOs opposed to whaling would probably take great umbrage at the latter comment. But there is at least one aspect of the dispute that is related to economics: the utilization of whales for whale-watching.

5.2.1 Whaling versus whale-watching

There is also a conflict over the use of whales as a resource. Whale-watching is a billion-dollar international industry involving more than 80 countries worldwide and is currently far more lucrative than whaling even within countries that conduct whaling (Hoyt, 2001; Parsons et al., 2003b). However, although it is economically valuable, there is concern that in many locations whale-watching may not be conducted sustainably and that the activity may have negative impacts on cetacean populations (Parsons, 2012).

Figure 5.2 A meeting of the scientific committee of the International Whaling Commission (2015, Bled, Slovenia—photo by E. C. M. Parsons)

Some countries at the IWC argue that as whale-watching does occur in countries that hunt whales, the two activities are compatible and can exist side by side (International Whaling Commission, 2008). However, others argue that whaling may be in conflict with whale-watching (Parsons and Rawles, 2003; Higham and Lusseau, 2007, 2008; Parsons and Draheim, 2009); for example, in Iceland, tourism revenue immediately decreased when whaling was introduced (Williams, 2006); tourists have expressed in several surveys that they would be less likely to visit countries that hunted whales (Parsons and Rawles, 2003; Parsons and Draheim, 2009); and although whale-watching occurs successfully in other parts of Japan, attempts at introducing whale-watching in Taiji (the location of dolphin drive hunts made famous by the Oscar-winning movie "The Cove") failed because of "discord" between tourists and dolphin hunters "resulting from different opinions over animal welfare" (Endo and Yamao, 2007, p. 180). Conflict has been particularly exacerbated when whale-watching tourists have witnessed whaling take place (International Fund for Animal Welfare, 2003; Berglund, 2006; Turley, 2007), which is perhaps unsurprising, as whale-watching tourists tend to be highly environmentally motivated and concerned about animal welfare and conservation issues (Lück, 2003; Parsons et al., 2003a; Rawles and Parsons, 2004).

The past and continuing conflict over whaling versus whale-watching, that is, whether the two are (or are not) compatible, which is the best use of whales as an economic resource, and what is the economic impact of whaling on whale-watching, are disputes unto themselves; nonetheless, the history of the debate over whale-watching also reveals an underlying conflict (see Section 5.3) which is arguably an identity-level conflict (see Section 5.4), as the clashing parties have ostensibly different fundamental values as to how whales should be utilized (non-consumptively/non-lethally via whale-watching vs. consumptively/lethally via whaling).

5.2.2 Is the argument about scientific whaling really about science?

People from both sides of the controversy have acknowledged that the conflict over so-called scientific whaling is in fact only partially over this dispute and that the scientific argument is just a superficial part of the conflict. For example, Morishita (2006) went on to point out additional, and deeper-rooted, areas of conflict (also see Comment):

> the whaling issue is often explained as the cultural and ethical collision between nations that regard whales as food and nations that see whales as special or even sacred. Political aspects also play an important role as it is politically important for the politicians in developed western nations to be seen as environmentally conscious by opposing whaling. (p. 802)

Clapham et al. (2007) responded to these statements but also recognized that the argument over the validity of scientific whaling went beyond the science:

> For many people in this debate, the issue is not that some whales are not abundant, but that the whaling industry cannot be trusted to regulate itself or to honestly assess the status of potentially exploitable populations. This suspicion has its origin in Japan's poor use of science, its often implausible stock assessments, its insistence that culling is an appropriate way to manage marine mammal populations, and its relatively recent falsification of whaling and fisheries catch data combined with a refusal to accept true transparency in catch and market monitoring. (p. 314)

Morishita (2006) and Clapham et al. (2007) touch on some of the deeper-rooted conflicts with respect to the scientific whaling dispute, which I will now describe further.

Comment

Differing cultural interpretations of marine mammals can cause and drive conflict, as this international example shows. This can also occur on a domestic, local level, however. For an example of this as it relates to Hawaiian spinner dolphins, see Chapter 8.

—*The Editors*

5.3 The underlying conflict

In the context of conflict theory, the "underlying conflict" is a history of unresolved disputes over what has happened in the past (Madden and McQuinn, 2014). At the IWC, there is a history of long and acrimonious arguments and personal attacks from both sides; broken promises; name-calling; and substantial financial and personal investment, with unsatisfactory results for all parties. There are, however, some specific issues which underlie the current dispute and which are worth highlighting.

5.3.1 Fisheries development aid

In the past couple of decades, many small developing countries have joined the IWC, and many of these countries have a pro-whaling stance at the IWC (e.g., Dominica, Gabon, Grenada, Morocco, Nicaragua, Palau, Solomon Islands, etc.). In some cases, there have been suggestions that Japan has employed foreign aid money to encourage, even coerce, these countries into voting for whaling. For example, an Antigua and Barbuda government newsletter stated that a $17 million fisheries grant from Japan was "as a direct result of its pro-whaling stance" and the prime minister of Antigua and Barbuda said "Quite frankly, I make no bones about it—if we are able to support the Japanese and the quid pro quo is that they are going to give us assistance" (Earth Island Institute, 2004). In 2000 the Dominican Minister for Environment, Agriculture and Fisheries resigned, stating that pressure from the Japanese government with regards to voting with them at the IWC was a major factor in his decision and furthermore that "[Japan] announced that if they couldn't get Dominica to come along with them [at the IWC] they would have to place Dominican [aid] projects under review" (Earth Island Institute, 2004). In 1987, a Japanese fisheries official stated: "When the Japanese Government selects the countries to which it provides fisheries grants, criteria include that the recipient country must have a fisheries agreement with Japan *and it must take a supportive position in various international organisations*" (emphasis added; Overseas Fisheries Cooperation Foundation, 1987). Moreover, the policy was admitted by Japanese government official and IWC delegate Maseyuku Komatsu during a TV interview, saying that the Japanese government had to use the "tools of diplomatic communications and promises of overseas development aid to influence members of the International Whaling Commission" (Watts, 2001).

Investigative journalists from the UK *Sunday Times* obtained documentation that incentives from Japan went beyond fisheries aid and included trips and daily allowances. The IWC Commissioner for Tanzania, Geoffrey Nanyaro, admitted to reporters that even prostitutes were supplied to fisheries officials (Stevenson, 2010).

Miller and Dolšak (2007) investigated recipients of Japanese fisheries aid funding (1999–2004) and found that there was a statistical link between receiving Japanese aid funds and voting a pro-whaling stance at the IWC. Stand and Tuman (2012) conducted a similar analysis and found that small member nations whose votes aligned with Japan at the IWC were statistically significantly more likely to receive Japanese aid funding.

The Japanese government has stated that it considers linking votes to fisheries aid to be standard diplomatic practice and points out that anti-whaling nations also frequently use economic leverage to gain assistance in treaty meetings or to curry diplomatic favor. Moreover, the fisheries aid provides benefits to some highly impoverished coastal communities in developing countries. The anti-whaling nations at the IWC largely see this as "vote buying" and express concern that it smacks of corruption and undermines the integrity of the treaty. As a result, there is a continued lack of trust and an escalation of conflict due to this issue.

Having said this however, there have been rumors of NGOs paying expenses for delegations from small developing countries to attend the IWC (and even paying the dues for those government parties) for representatives to attend the meeting on behalf of the anti-whaling lobby (Sakaguchi, 2013). However, these claims have not been substantiated. Even if these practices did occur, given that even the largest NGOs have limited budgets, the money involved would be considerably smaller than the amounts discussed above.

5.3.2 Marine resource treaties

The IWC is considered by many to be a fishery management treaty rather than a conservation treaty, although it deals with marine mammals. Much of the science associated with the IWC is fisheries science, and the IWC commonly uses fisheries management concepts such as maximum sustainable yields (Mace, 2001). Many of the same scientists and delegation members at the IWC participate in other marine resource treaty organizations such as the International Commission for the Conservation of Atlantic Tunas, the Inter American Tropical Tuna Commission, and the Convention for the Conservation of Antarctic Marine Living Resources. Negotiations in these international meetings have much more economic importance than whaling—the tuna fishery in the Pacific alone is a multibillion dollar industry in which Japan is a major player (e.g., Williams and Terawasi, 2010). The two sides at the IWC have a history of using whaling as a bargaining chip in negotiations at other forums, and debates at the IWC may actually be rooted in marine resource use clashes in these other forums (Clapham et al., 2007). For example, proposals for takes of charismatic whales such as humpbacks (which may be politically important in anti-whaling nations) might be bartered in order to gain leverage in negotiations over more financially important tuna quotas. A history of using whales as a bargaining chip on the larger political stage of global fisheries negotiations is an underlying conflict that further fuels reactivity and entrenchment of positions at the IWC.

5.3.3 Diplomatic ties that bind

Strong political ties exist between Japan and many anti-whaling nations, economically and militarily. For example, Australia, New Zealand, Europe, and the U.S. are major trading partners with Japan; and the U.S., UK, and Japan were three of the most powerful nations in the coalition of countries in the recent Iraq War, in which Saddam Hussein's regime was toppled. The IWC may be one of the few forums where Japan can be independent of these ties and "flex its political muscles," as whaling is such a minor issue compared to these economic and military activities. Conversely, it's possible that the anti-whaling nations may feel they must "play nice" with Japan at the IWC so as not to jeopardize support in other forums and so have not implemented major diplomatic actions (e.g., sanctions) to curb scientific whaling in recent years.

5.4 Identity-level conflict

"Identity-level" conflict entails disputing parties having prejudices and assumptions about the parties involved based on the groups to which they belong (Madden and McQuinn, 2014). In the IWC context, there are frequently gross prejudices and stereotypes at play; for example, pro-whaling nations are labeled as being "heartless whale killers," whereas pro-conservation countries are considered "unrealistic bunny huggers." This last level also includes assumed offenses to national sovereignty and pride, which is a major factor exacerbating conflict at the IWC (Mitchell, 1998).

5.4.1 Public values and identity

There are certainly basic identity-level conflicts over attitudes to whales. For example, in many of the anti-whaling nations, whales are considered special aesthetically, culturally, or both (e.g., Australia, New Zealand, and Europe). Objections to whaling are often based on moral grounds in these countries, that is, whaling is inhumane or just "wrong" because of the "specialness" of whales. Hamazaki and Tanno (2001) noted that concern for whale conservation seemed to be correlated with increasing disapproval of whaling. Conservation of whales in these anti-whaling countries may have a high priority, regardless of the actual conservation status of the animals. For example, a survey in one part of the UK found that support was strong for cetacean conservation (Scott and Parsons, 2005; Howard and Parsons, 2006), to the extent that 40% of those surveyed said that they would be more likely to vote for a politician if they supported whale conservation (Howard and Parsons, 2006). Several studies have found a majority of the public in anti-whaling nations disagree with whaling, and if anything this public opposition and disapproval is increasing (Freeman and Kellert, 1994; Kellert, 1999; Hamazaki and Tanno, 2001; Scott and Parsons, 2005). This is arguably increasing the polarization among certain countries regarding their stances toward whaling and, thus, increasing the identity-level conflict.

In comparison, the attitudes of the public in pro-whaling nations are that whales are just another type of livestock or fish, that anti-whaling nations are being unreasonably speciesist in treating whales as somehow different from any other species that might be exploited (Kirby, 2001), and, unsurprisingly, that there is a need for continued whaling. In addition, they argue that there is support for the idea that whaling is a sustainable fishery (Freeman and Kellert, 1994; Kellert, 1999; Hamazaki and Tanno, 2001; Bowett and Hay, 2009).

5.4.2 Saving "face" and the Sea Shepherd factor

One identity-level factor that is often not understood or readily appreciated by westerners is the concept of "saving face" (put simply, avoiding embarrassment or a loss of prestige; an important social factor in several Asian cultures in particular). The Japanese government heavily invested historically, economically, and politically in the whaling industry; it would be difficult now for them to back down and stop whaling without a huge loss of face among their governmental colleagues, the Japanese public, and internationally. Also, dismissing data produced by Japan causes a loss of face for Japanese scientists, and arguments over the scientific validity of whaling under permit can lead to embarrassment for Japanese scientists; thus, scientists and government officials may become entrenched in defending their position.

It is this cultural factor that makes it unlikely that the activities of the NGO Sea Shepherd Conservation Society would do anything more than further entrench the Japanese government in their support of Antarctic whaling. Sea Shepherd shadows and tries to prevent Japanese whaling vessels from killing whales in Antarctica (see <http://www.seashepherd.org/whales>). If Japan ceased scientific whaling in the Southern Ocean, Sea Shepherd would undoubtedly claim this as a victory. This would, however, place Japan in the position of appearing as though they—a major world power—had been defeated by a small activist group labeled by many in their government as terrorists (Perry, 2013). This would lead to a loss of face for the Japanese government and potentially cause a huge dent in nationalistic pride. The specter of this has thus increased the resolve of the Japanese government to continue whaling and further entrenched them in their position. Even if Japan wanted to stop whaling in Antarctica (because, for example, it may be not lucrative), the situation with Sea Shepherd makes it difficult for them to do so without a major loss of face.

5.4.3 Other identity-level conflicts

Whaling and the IWC is a complex, nuanced situation, and it is likely that there are many other identity-level conflicts at play (these are summarized in Table 5.1). Some of these conflicts may have originally been underlying conflicts, but because they have existed for so long and have become so institutionalized, they are now effectively identity-level conflicts.

Whaling is a high profile issue that attracts a lot of media attention, and government officials defending their pro- or anti-whaling stances at the IWC can "win votes" back home by appealing to politicians' home bases and to conflicts and concerns at the identity level—whether speaking to nationalistic Japanese citizens or environmentally motivated or "animal-loving" voters. This strengthens identity-level conflict, exacerbates the polarization, and further impedes the ability to find agreement.

Much of the criticism of whaling activities focuses on the Japanese government. There is less focus on the whaling activities of Norway and Iceland, even though the latter has also conducted scientific whaling and now conducts a very controversial—and arguably illegal—commercial whaling program (Parsons et al., 2012). Because there is less focus on the activities of Iceland and Norway, this has been perceived and portrayed by Japanese officials as imperialist or racist (Kirby, 2001), setting the stage for another layer of potential identity-level conflict. Anti-whaling advocates argue that Japanese whaling is larger in scale and scope than Icelandic whaling, that the activities of Norway and Iceland are under reservations to the Convention, and that they are not being conducted in, and therefore undermining, the Southern Ocean sanctuary.

Table 5.1 Other identity-level conflicts

Coastal Japanese communities vs. anti-whalers	Those in Japanese coastal communities argue that whaling is part of a long-standing cultural tradition, akin to indigenous or aboriginal whaling in locations such as Alaska. Anti-whaling nations are therefore opposing their cultural identity.
Anti-whalers vs. fishermen	Fishermen believe anti-whalers place whales above their families and livelihoods, either by denying them the right to catch whales or by supporting a predator of "their" fish stocks. Whales taken in the JARPN program are in the Japanese Exclusive Economic Zone and by rights belong to Japanese fishermen. Therefore, the anti-whaling stance is a threat to their way of life as fishermen.
Axis vs. Allies	Many of the anti-whaling countries were opponents of Japan in WWII. Some still feel resentment toward historical enemies, and their treatment after hostilities ended.
Imperialists vs. the colonized	Japan (i.e., post-WWII) and its allies (e.g., African and Caribbean states) have been occupied historically by many of the anti-whaling countries. There is still long-term/historical resentment because of this.
Developing nations vs. developed	Although Japan is developed, many of its allies are developing countries, and the identity-level conflict can be viewed as conflict and resentment between the *haves* and *have-nots*.

5.5 The "deal"

In 2010 there was an attempt to broker a deal on whaling at the IWC meeting in Morocco. The United States and New Zealand developed a compromise package that would allow Japan to continue whaling, but with potentially smaller catch sizes and some oversight (Jolly, 2010). Part of the argument for the deal was that the IWC might collapse under the unresolved conflict and deadlock. Arguably the "deal" was agreeable in purely scientific terms, as catch sizes were likely to be sustainable based on current estimations of whale stock sizes, or at least more so than current catch levels via scientific whaling. Moreover, the deal would ban whaling activities in the Southern Ocean (Jolly, 2010). However, the initiative ultimately failed, in large part because other, more deeply entrenched levels of conflict were not considered or addressed. For example, many nations were concerned that Japan would not adhere to the deal in good faith, would "top up" their commercial quota with scientific whaling, and that there was a lack of oversight and enforcement mechanisms. There was a widespread lack of trust in Japan's intentions, perhaps based on past experiences and interactions, but possibly also prejudices (personal observation). Arguably the divisive status quo was preferable to many nations, and an actual collapse of the IWC was deemed unlikely (Iliff, 2008). A petition of more than 200 scientists and experts was drafted and circulated while discussions over the deal were going on, which may have given some countries an "out" based on dispute-level scientific concerns (Parsons et al., 2012), although the actual concerns and conflicts were more at an underlying or identity level. An additional identity-level conflict may have been that in 2010, the United States (the leader of this initiative) was facing anti-U.S. sentiment, and a lack of trust in U.S. leadership was running high in Europe, the result of inquiries and disclosures over the Iraq War.

During the brokering of the deal, diplomatic outreach to Japan by the U.S. IWC Commissioner Monica Medina offered to have Sea Shepherd inconvenienced and potentially financially impacted by having their tax-free charitable status under U.S. law reviewed—and possibly revoked—by the Internal Revenue Service (IRS). A summary of this communication was released to Wikileaks (Koh, 2011). As a result, there is now an unsurprising lack of trust between Sea Shepherd and the U.S. government (developing another underlying conflict), making it less likely that negotiations between anti-whaling nations and Sea Shepherd would be successful.

5.6 Conclusions

The Japanese government may be more sensitive to and engage with the underlying/identity-level conflicts more than anti-whaling countries. Officials of the latter often deal with whaling on a purely scientific basis and don't acknowledge the underlying and identity-level conflicts, even their own. Japanese officials, however, use identity to fuel support for their campaigns—for example producing outreach materials for fishermen in developing countries portraying whales as animals that eat "their fish." By primarily using science to try to address the opposition, the anti-whaling nations are missing the point of what is driving the conflict. Underlying and identity-level conflict, not scientific disputes, are ultimately the cause of the log jam at the IWC, and these have to be assessed, acknowledged, and transformed if the dispute is to be sustainably settled. If we don't, further deals, compromises, or "ways forward" will likely collapse and potentially exacerbate the social conflict by adding yet another antagonistic failure.

5.6.1 Are there solutions?

Instead of getting mired down in the scientific aspects of the IWC debate, there needs to be more recognition of the existence of real underlying and identity-level conflicts at the IWC by those who want to settle the dispute over scientific whaling, along with a more rigorous analysis of these conflicts (i.e., less biology, more psychology). With such a complicated deadlock, perhaps there should be change in the discussion framework.

Some propose changing the argument from a deadlocked scientific one about whale numbers to an economic dispute—whaling is arguably not economically viable—and putting the arguments about whaling into a purely economic arena might change the historical landscape of the dispute. However, economic discussions that fail to address the underlying and identity-level conflicts will also fail to address what is really driving the conflict (Tinch and

Phang, 2009; Madden and McQuinn, 2014). Japan is unable to sell the whalemeat it already harvests (Tinch and Phang, 2009), and now the Japanese market is also being flooded with fin whale meat imported from Iceland. In addition, recent International Maritime Organization regulations for the Southern Ocean have prohibited vessels using certain types of fuel (MAR-POL Annex I; Chapter 9) and with certain types of hulls (International Maritime Organization, 2010), which means that Japan's main whaling ships may not be able to operate in these waters for much longer, as they are noncompliant with these regulations and guidelines. The recent ruling by the International Court of Justice (see textbox on the 2014 IWC meeting below) and the subsequent resolution on scientific whaling by the IWC adds another obstacle to Antarctic whaling for Japan.

One option is to "give Japan an out," that is, to allow a graceful retreat from whaling or enable them to change their stance at the IWC without losing (too much) face. With many of the IWC scientists and delegates who participated during some of the more acrimonious years of whaling, both commercial and scientific, being replaced by younger delegates and scientists with less historical "baggage," some (but not all) of the underlying and identity-level conflicts may dissipate as new participants may not identify themselves with some of the ideologies of the "old guard." Regardless, constructing a scenario where the Japanese government and scientists can engage in such a way that they are not publically embarrassed or suffer a loss of dignity is crucial.

A worst case scenario, if the IWC is so polarized and entrenched in conflict, might be to either recognize another international treaty entity (such as the Convention for Migratory Species) as the competent authority for whale management or even to develop a new international treaty organization and start again from scratch. Many countries, however, have become very invested in the IWC and so want it to continue, as they are wary that a new body may have its own problems. Moreover, there is nothing to prevent a new body from becoming quickly populated with IWC "old guard" should a shift be made, bringing along with them their previous conflict baggage. Regardless, more needs to be done to resolve the underlying conflicts and to reconcile the identity-level conflict, as if this is not done, the current tactic of a purely scientific debate on the value scientific whaling will fail to bring a resolution.

Lessons learned

- Clarity when it comes to laws, policies, and treaties is vital. When different parties have different interpretations of the same laws and agreements, conflict can ensue.
- Conflict over wildlife might be used as a bargaining chip in larger conflicts (consciously or not). Look for what other interactions are happening between groups in conflict with each other (even those having nothing to do with wildlife) to see what else might be playing a role in your situation.
- It is imperative to understand cultural differences between different sides of a conflict, as cultural misunderstandings can quickly result in barriers to productive discussion.

—The Editors

References

Berglund, N. (2006). "Whale shot in front of tourists." *Aftenposten*, July 4, 2006, <http://www.aftenposten.no/english/local/article1376980.ece/>, accessed August 10, 2008.

Bowett, J. and Hay, P. (2009). Whaling and its controversies: examining the attitudes of Japan's youth. *Marine Policy*, **33**(5):775-83.

Burns, W. C. (1997). The International Whaling Commission and the future of cetaceans: problems and prospects. *Colorado Journal of International Environmental Law and Policy*, **8**(1):31-88.

Clapham, P. J., Berggren, P., Childerhouse, S., et al. (2003). Whaling as science. *Bioscience*, **53**(3):210-12.

Clapham, P. J., Childerhouse, S., Gales, N. J., et al. (2007). The whaling issue: conservation, confusion, and casuistry. *Marine Policy*, **31**(3):314-19.

Corkeron, P. J. (2009). Reconsidering the science of scientific whaling. *Marine Ecology Progress Series*, **375**:305-9.

Earth Island Institute. (2004). "Japan denies vote-buying, but vote-sellers say yes." *eCO* 2004, 56(4), <http://www.earthisland.org/immp/eco2004_issue4.html>, accessed June 23, 2013.

Ellis, R. (1991). *Men and Whales*. New York: Knopf.

Endo, A. and Yamao, M. (2007). Policies governing the distribution of by-products from scientific and small-scale coastal whaling in Japan. *Marine Policy*, **31**(2):169-81.

Environmental Policy and Law. (2002). Iceland readmitted to commission. *Environmental Policy and Law*, **32**(6):256-7.

Freeman, M. M. R. and Kellert, S. R. (1994). International attitudes to whales, whaling and the use of whale products: a six country survey. In Freeman, M. M. R. and Kreuter, U. P. (eds), *Elephants and Whales: Resources for Whom?* London: Gordon and Breach, pp. 293-303.

Gales, N. J., Kasuya, T., Clapham, P. J., and Brownell, R. L. (2005). Japan's whaling plan under scrutiny. *Nature*, **435**(7044):883-4.

Hamazaki, T. and Tanno, D. (2001). Approval of whaling and whaling-related beliefs: public opinion in whaling and nonwhaling countries. *Human Dimensions of Wildlife*, **6**(2):131-44.

Hatanaka, H. (2005). Answering the critics of Japanese whale research. *Nature*, **436**(7053):912.

Higham, J. E. S. and Lusseau, D. (2007). Urgent need for empirical research into whaling and whale-watching. *Conservation Biology*, **21**(2):554-8.

Higham, J. E. S. and Lusseau, D. (2008). Slaughtering the goose that lays the golden egg: are whaling and whale-watching mutually exclusive? *Current Issues in Tourism*, **11**(1):63-74.

Howard, C. and Parsons, E. C. M. (2006). Attitudes of Scottish city inhabitants to cetacean conservation. *Biodiversity and Conservation*, **15**(14):4335-56.

Hoyt, E. (2001). *Whale-Watching 2000: Worldwide Tourism Numbers, Expenditures, and Expanding Socioeconomic Benefits*. Crowborough: International Fund for Animal Welfare.

Iliff, M. (2008). Compromise in the IWC: is it possible or desirable? *Marine Policy*, **32**(6):997-1003.

International Fund for Animal Welfare. (2003). *Iceland Kills Whale in Whale-Watching Bay*. <http://www.ifaw.org/ifaw/general/default.aspx?oid1/471311/>, accessed August 10, 2008.

International Maritime Organization. (2010). *Guidelines for Ships Operating in Polar Waters*. London: International Maritime Organization.

International Whaling Commission. (2008). Whale-watching. In *Annual Report of the International Whaling Commission 2007*. Cambridge: International Whaling Commission, pp. 45-7.

International Whaling Commission. (2013). *Scientific Permit Whaling*. <http://iwc.int/permits>, accessed June 23, 2013.

Jolly, D. (2010). "Under pressure, Commission discusses lifting whaling ban." *The New York Times*, June 21, 2010, <http://www.nytimes.com/2010/06/22/world/22whale.html?_r=1&>, accessed June 23, 2013.

Kellert, S. R. (1999). *American Perceptions of Marine Mammals and Their Management*. Washington, DC: Humane Society of the United States.

Kirby, A. (2001). "Whaling 'safe for a century." *BBC News*, October 4, 2001, <http://news.bbc.co.uk/2/hi/science/nature/1578812.stm>, accessed June 23, 2013.

Koh, Y. (2011). "Wikileaks Japan: whale diplomacy." *Wall Street Journal*, January 3, 2011, <http://blogs.wsj.com/japanrealtime/2011/01/03/wikileaks-japan-whale-diplomacy/>, accessed June 23, 2013.

Lück, M. (2003). *Environmentalism and On-tour Experiences on Wildlife Watch Tours in New Zealand: A Study of Visitors Watching and/or Swimming with Dolphins*. Ph.D. thesis. Dunedin: University of Otago.

Mace, P. M. (2001). A new role for MSY in single-species and ecosystem approaches to fisheries stock assessment and management. *Fish and Fisheries*, **2**(1):2–32.

MacLeod, D. A. (1994). International consequences of Norway's decision to allow the resumption of limited commercial whaling, *International Legal Perspectives*, **6**(1):131–58.

Madden, F. and McQuinn, B. (2014). Conservation's blind spot: the case for conflict transformation in wildlife conservation. *Biological Conservation*, **178**:97–106.

Miller, A. R. and Dolšak, N. (2007). Issue linkages in international environmental policy: the International Whaling Commission and Japanese development aid. *Global Environmental Politics*, **7**(1):69–96.

Mitchell, R. B. (1998). Discourse and sovereignty: interests, science, and morality in the regulation of whaling. *Global Governance*, **4**:275–93.

Morishita, J. (2006). Multiple analysis of the whaling issue: understanding the dispute by a matrix. *Marine Policy*, **30**(6):802–8.

Overseas Fisheries Cooperation Foundation. (1987). *Conference Report: Symposium on South Pacific Fisheries Development*. Tokyo: Overseas Fisheries Cooperation Foundation.

Parsons, E. C. M. (2012). The negative impacts of whale-watching. *Journal of Marine Biology*, 2012: Article ID 807294.

Parsons, E. C. M., Bauer, A., McCafferty, D., Simmonds, M. P., and Wright, A. J. (2012). *An Introduction to Marine Mammal Biology and Conservation*. Sudbury, MA: Jones & Bartlett Publishing.

Parsons, E. C. M. and Draheim, M. (2009). A reason not to support whaling: a case study from the Dominican Republic. *Current Issues in Tourism*, **12**(4):397–403.

Parsons, E. C. M. and Rawles, C. (2003). The resumption of whaling by Iceland and the potential negative impact in the Icelandic whale-watching market. *Current Issues in Tourism*, **6**(5):444–8.

Parsons, E. C. M., Warburton, C. A., Woods-Ballard, A., et al. (2003a). Whale-watching tourists in West Scotland. *Journal of Ecotourism*, **2**(2):93–113.

Parsons, E. C. M., Warburton, C. A.,Woods-Ballard, A., Hughes, A., and Johnston, P. (2003b). The value of conserving whales: the impacts of cetacean-related tourism on the economy of rural West Scotland. *Aquatic Conservation*, **13**(5):397–415.

Perry, N. (2013). "Sea Shepherd claims 'false': Japan." *The Australian*, March 28, 2013, <http://www.theaustralian.com.au/news/breaking-news/sea-shepherd-claims-false-japan/story-fn3dxiwe-1226608349355>, accessed June 23, 2013.

Rawles, C. J. G. and Parsons, E. C. M. (2004). Environmental motivation of whale-watching tourists in Scotland. *Tourism in Marine Environments*, **1**(2):129–32.

Reilly, S. B., Bannister, J. L., Best, P. B., et al. (2008a). *Balaenoptera physalus*. In IUCN 2012, *IUCN Red List of Threatened Species. Version 2012.2*. <http://www.iucnredlist.org>, accessed June 23, 2013.

Reilly, S. B., Bannister, J. L., Best, P. B., et al. (2008b). *Balaenoptera bonaerensis*. In IUCN 2012, *IUCN Red List of Threatened Species. Version 2012.2*. <http://www.iucnredlist.org>, accessed June 23, 2013.

Sakaguchi, I. (2013). The roles of activist NGOs in the development and transformation of IWC regime: the interaction of norms and power. *Journal of Environmental Studies and Sciences*, **3**(2):194–208.

Scott, N. J. and Parsons, E. C. M. (2005). A survey of public opinions in Southwest Scotland on cetacean conservation issues. *Aquatic Conservation*, **15**(3):299–312.

Scott, S. (1999). Australian diplomacy opposing Japanese Antarctic whaling 1945–1951: the role of legal argument. *Australian Journal of International Affairs*, **53**(2):179–92.

Stevenson, A. (2010). "Flights, girls and cash buy Japan whaling votes." *The Sunday Times*, June 13, 2010, <http://www.timesonline.co.uk/tol/news/uk/article7149086.ece>, accessed June 23, 2013.

Strand, J. R. and Tuman, J. P. (2012). Foreign aid and voting behavior in an international organization: the case of Japan and the International Whaling Commission. *Foreign Policy Analysis*, **8**(4):409–30.

Tinch, R. and Phang, Z. (2009). *Sink or Swim: The Economics of Whaling Today*. <http://www.wdcs.org/submissions_bin/economics_whaling_report.pdf>, accessed June 23, 2013.

Turley, J. (2007). *Japanese Version of Whale-Watching: Whalers Kill a Whale in Front of Eco-Tourists*. <http://jonathanturley.org/2007/08/27/japanese-version-ofwhale-watching-whalers-kill-a-whale-in-front-of-eco-tourists/>, accessed August 10, 2008.

Wansbrough, T. (2004). On the issue of scientific whaling: does the majority rule? *Review of European Community and International Environmental Law*, **13**(3):333–9.

Watts, J. (2001). "Japan admits buying allies on whaling." *The Guardian*, July 18, 2001, <http://www.guardian.co.uk/world/2001/jul/19/japan.whaling>, accessed June 23, 2013.

Williams, N. (2006). Iceland shunned over whale hunting. *Current Biology*, **16**(23):R975–6.

Williams, P. and Terawasi, P. (2010). *Overview of Tuna Fisheries in the Western and Central Pacific Oceanic, Including Economic Conditions–2009*. <http://www.wcpfc.int/node/2911>, accessed June 23, 2013.

6

Conflict and Collaboration in Marine Conservation Work

Transcending Boundaries and Encountering Flamingos

Sarah Wise

6.1 Introduction

In the Bahamas, and elsewhere, Marine Protected Areas (MPAs) can be sources for conflict. Protected areas are often introduced as solutions to declining fisheries, habitat destruction, or vulnerable biodiversity. Management systems such as enclosures are put in place to protect resources and limit damage to valuable ecosystems and species. They are specialized spaces with restricted access for specific purposes. As such, protected areas are perceived very differently according to different ideas about the role and value of nature, historical experiences within a given environment, and cultural paradigms. They are social processes in of themselves (Walley, 2004) and therefore are dynamic areas of human and interspecies interaction. Just as these interactions can provide opportunity for conflict, they also allow for the possibility of collaboration, resolution, and greater understanding. Using the West Side National Park (WNP) expansion project in Andros Island, Bahamas, as a case study, this chapter explores the levels of conflict associated with disputes surrounding resource access and tenure systems in order to better understand the challenges of park implementation and marine conservation.

Marine conservation in the Bahamas relies on strong public support and investment in conservation goals. To facilitate this investment and engender support, often iconic species are used to lead the charge for conservation. In the Bahamas it is the West Indian flamingo (*Phoenicopterus ruber*; Figure 6.1) that best represents conservation in the archipelago nation. The West Indian flamingo, with its brilliant color and ungainly flying posture, has become an icon of conservation in the Bahamas. The flamingo is the Bahamas state bird and has been at the heart of the conservation debate since the 1950s, when The National Audubon Society, in response to international pressure to protect the West Indian flamingo, conceptualized, spearheaded, and funded the Bahamas National Trust (BNT) as a regional conservation organization. The flamingo flies across the BNT's coat of arms, and its image is featured

Human–Wildlife Conflict: Complexity in the Marine Environment. Edited by Megan M. Draheim, Francine Madden, Julie-Beth McCarthy, and E. C. M. Parsons © Oxford University Press 2015. Published 2015 by Oxford University Press.

prominently on the promotional literature for conservation projects throughout the islands. In Andros Island, the flamingo takes on greater meaning, because the west side of the island is home to a non-breeding flock ranging from 300 to over 1,000 birds.

To protect the resting flock of West Indian flamingo in Andros, a group of local conservation organizations created the WNP in 2002. Three years later, the same group led a move to expand the park to include other valuable species and habitat. While few people live on the west side of Andros, the coastline is heavily used by sport and commercial fishers, hunters, spongers, and straw collectors for traditional basket-weaving. The park expansion proposal sparked conflict among resource users, conservationists, land owners, and residents. This chapter provides an overview of the WNP expansion plan and the community responses. It then goes on to describe the conflicts surrounding issues of rightful resource tenure and belonging. Using the levels of conflict model (Chapter 1; Madden and McQuinn, 2014), the conflict surrounding WNP expansion is analyzed. Finally, a moment of exceptional beauty is detailed, when boundaries appear to blur, perhaps even vanish, during a moment of engagement with an iconic species under protection.

6.1.1 The WNP

In 2002 a group of local conservationists created a small national park on the largely uninhabited west side of Andros to conserve "exceptional natural systems in Andros" as well as preserve the resting flock of West Indian flamingos (IUCN, 2014). Three years later, the

Figure 6.1 *Phoenicopterus ruber.* Photo Credit: By Virginia Ortiz (Own work) [CC BY-SA 3.0 (http://creativecommons.org/licenses/by-sa/3.0)], via Wikimedia Commons.

conservation alliance grew, joining forces with international NGOs in order to expand the original park. While boundary lines were not yet established, the conservation alliance's goal was to protect as much "pristine wilderness" (The Nature Conservancy, 2006a) as possible. The west side of Andros is a vast expanse of shallow seas, mud flats, and shrub palms. The area is known for its rich natural resources, stark beauty, and the resident flock of West Indian flamingoes. Significant outreach efforts were conducted by Bahamian and international conservation organizations, and a Rapid Ecological Assessment (REA) was performed on the west side in order to determine high-priority areas for conservation. There were numerous town meetings, school presentations, site visits, workshops, and community events. The local government was notified and asked to support the proposal. An interdisciplinary team of Bahamian researchers investigated community response to the park proposal and to the idea of marine conservation in general. Other researchers surveyed households as well as targeted resource users on topics relating to the WNP such as livelihoods, ecological knowledge and resource use, land ownership, and historical ties to the area. Tremendous resources were directed to the REA, as well as outreach efforts which lasted from 2006 until 2009.

The plan was to create the park and establish boundaries later. One conservation agent explained the reasoning for promoting the expansion plan despite the conflict among stakeholders: "We need to strike while the iron is hot. You can't please everyone. You know Androsians, they're going to complain no matter what. We don't need more town meetings to know that" (Wise, unpublished observation from an interview conducted in October 2009). Following the vague directive of "the protected area planning should take the approach of protecting representative habitat types from throughout western Andros" (Bahamas National Trust, 2012, p. xiii), the push was simply to expand the original park north- and southward to include as much ecologically valuable habitat as possible. In 2009 the Bahamian government formalized the expansion plan to dramatically expand the park from 300,000 acres to 1.3 million acres, in effect enclosing much of the western length of Andros Island's coastline for the purpose of vulnerable species and habitat protection (Bahamas National Trust, 2012; Figure 6.2).

The newly proposed WNP was an area roughly 80 miles long and 25 miles wide, one of the largest protected areas in the Western Atlantic/Caribbean region (Bahamas National Trust, 2012). The management plan describes the park as regulated through a series of management zones, areas zoned for specific uses such as conservation, sensitive resource use, camping, and day use (Bahamas National Trust, 2012). The conservation zone encompasses the entire park and is aimed at conserving natural resources and processes while "accommodating uses and experiences that do not adversely affect the ecological integrity or the scenic quality and serenity of the area" (Bahamas National Trust, 2012, p. 20). No development is allowed beyond minimal infrastructure to support the conservation goals.

6.1.2 Community response

During outreach meetings, conservation managers were firm in their assertion that sustainable resource use would continue and livelihoods would be minimally impacted by the park

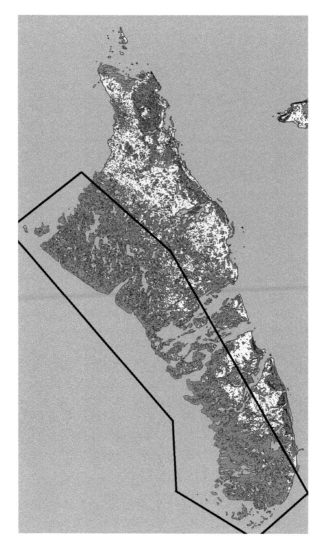

Figure 6.2 Adapted topographic map of Andros Island, Bahamas, shaded at 15 ft. (4.572 m) contour intervals; West Side National Park proposed boundaries in black. Credit: By Lithium6ion (Own work) [Public domain], via Wikimedia Commons, adapted by author.

expansion; however, community responses remained mixed. Interviews with residents conducted in 2006–2009 indicated that many people believed the conservation effort was led by a small group of social elite. These individuals were frequently associated in the Bahamas, and in Andros specifically, with attempts to claim areas of land and sea in the name of conservation but for personal benefit. The size of the proposed expansion took many Androsians by surprise, and soon disputes erupted over park boundaries, restricted access to resources, and existing tenure claims on the west side of Andros. While there was generalized support for protecting important fisheries from overharvesting and foreign development, there was suspicion and unease over the idea of enclosing the west coast of Andros for the sake of conservation. These tensions were tied to recent conflicts over access to resources and land claims as well as a long history of colonial rule and racial injustice. Conservation organizations promoted the WNP park system as a necessary move for the health of the environment; however, many Androsians saw the park placement and boundary lines as simply extending certain private property claims.

The resulting conflict was broad-ranging and profound. Participation in management meetings diminished, and conservation agents encountered resistance in many forms, including poaching, uprooted signage, and verbal confrontations. Some members of the conservation alliance withdrew their support as the conflict grew.

6.2 Background

6.2.1 Andros Island and land tenure

Andros is the largest island in the Bahamas, running 100 miles long and 50 miles at its widest point (Figure 6.3). The seas surrounding Andros have long been active sites of industry, whether harvesting sponge, aragonite, or lobster. The shallow sandy banks on the west harbor several commercial species, including spiny lobster (*Panulirus argus*), stone crab

Figure 6.3 Andros skyline. Credit: By GdML (Own work) [CC BY-SA 3.0 (http://creativecommons.org/licenses/by-sa/3.0)], via Wikimedia Commons.

(*Menippe mercenaria*), various sponge species, queen conch (*Strombus gigas*), and scalefish, including bonefish (*Albula vulpes*), a popular fish for the small but highly lucrative sports fishing industry on the island as well as for subsistence. International interests include commercial trawlers and long liners that fish illegally, absorbing the minimal costs on the rare occasion when a vessel is apprehended. The people of Andros rely heavily on the marine environment for subsistence. Fishers frequent the area for commercial and subsistence fishing; however, the west side is valued most highly for its extensive bonefish flats. Bonefish have fed Androsians for centuries. Foreign fishers pay thousands of dollars a week to travel to the west side of Andros to fish for bonefish, permit, and jack, generating $1.9 million annually for the nation (Fedler, 2010).

The west side is considered "Crown land," government land held in trust for the people of the Bahamas by the Bahamian Government. Crown land holds important meaning for many Bahamians as land that is available to citizens. In the 1700s, Loyalists arrived from the rich fertile grounds of Florida and the Carolinas, with aspirations to develop a strong agricultural economy supported by slave labor. Claiming large tracks of Crown land, they made several attempts at commercial farming, along the same lines as was done in other British colonial islands. The lack of space and arable soil led to the frequent failure of plantations. Bankrupt land owners fled to Nassau, leaving their property—both material and human—behind (Craton, 1962). Those people left behind settled the islands and built communities, claiming the land and sea through daily practice.

In many ways, the failure of the Bahamas as a plantation state allowed for greater independence of the enslaved and later of the working class (Johnson, 1991). After emancipation in 1834, the Bahamas underwent a series of land laws privileging existing landholders while attempting to formalize commonage property in order to provide newly freed slaves with some access to land. These laws released the white merchant class from any financial responsibility for their prior enslaved peoples and formally recognized customary tenure laws, which had a long history among enslaved Africans and their descendants in the Bahamas. According to historians, "The process of emancipation, when the black majority in the Bahamian population instantly changed from being property to owning it, though, was a crucial watershed in Bahamian landholding for the imperial government as well as for the former masters and slaves" (Craton and Saunders, 2000, p. 48). After emancipation, the demand for land grew. The slaves were freed and continued successfully to farm small plots of land for subsistence, building on shared notions of common property that remain relevant today (Craton and Saunders, 2000). This land became known as "Generation land" or "Family land": land owned and maintained over generations by black Bahamians.

6.2.2 History of enclosure conservation in the Bahamas

In 2009 the WNP expansion plan emerged from a broader move to manage natural resources through enclosure conservation. Protected areas have a long history and wide presence in the shape of parks, nature reserves, wildlife sanctuaries, and other protected terrestrial places

(Hulme and Murphree, 2001). In the Bahamas, MPAs date back to the creation of the Exuma Cays Land and Sea Park in 1956. In the early 2000s, the Bahamas returned to the model of enclosure conservation to protect valuable marine resources.

Studies comparing MPAs and non-protected areas suggest that there are greater species diversity and abundance within park boundaries (Polunin and Roberts, 1993; Roberts, 1995; Roberts et al., 2001; Halpern and Warner, 2002). In the Caribbean, research has indicated a greater density and biomass of larger grouper species in no-take areas (Chiappone et al., 2000). Protected areas have been widely promoted for marine systems as solutions to declining fisheries and marine degradation (Agardy, 1997, 2000; Botsford et al., 2003; Roberts et al., 2005). Critics argue, however, that conservation agents often simplify both the biological and the sociocultural aspects of protected area conservation (West et al., 2006; Wood et al., 2008; Campbell et al., 2009). Some social effects of enclosure conservation include restricted access to resources, shifts in land tenure, geographic and cultural displacement, and changes in the ways people view and use their environment and the world (Orlove and Brush, 1996; McCay, 2002; Mascia, 2003; West and Brockington, 2006; Wilson, 2007). Enclosure conservation is designed to control the fluid and ever-changing interactions between humans, wildlife, and their habitats, regulating human activity as well as the areas set aside for protection (Wise, 2014).

Enclosure has become the primary means of protection of the sea, marine resources, biodiversity, and Bahamian livelihoods. The pressure to increase protected areas in the country was strong after signing—along with several other Caribbean nations—onto the "2000 Caribbean Challenge," an initiative executed by The Nature Conservancy to protect 20% of the country's marine and coastal environments, and 10% of land by 2020. The initial protected areas in Andros were part of Bahamas' move to double the size of the country's national parks. Furthermore, the enclosed areas are progressively getting larger in scale. MPA proposals have expanded in scope and complexity from small regional MPAs to larger networks of linked protected areas. This context helps account for the move to expand the WNP from the small circle of land protecting the West Indian flamingo and surrounding habitat to a great swath of land and sea covering 1.3 million acres of the western coastline of Andros.

6.2.3 The conservation alliance

In 2002, The BNT formed a conservation alliance with Androsian conservation organizations to propose a series of protected areas on Andros Island, Bahamas. As part of that system, BNT named 300,000 acres of western low-lying coastline, which is frequented by the West Indian flamingo, as the WNP. As soon as the Bahamian government granted the proposal, the partnership began to develop a plan to expand the park.

Credited with building capacity for a park system on Andros, it was this alliance that received funding to conduct an REA of the valuable ecosystems in the west of Andros. The alliance was promoted as an example of an effective integration of international resources (e.g., money and international influence) with regional resources (labor and territorially based influence). Each partner in the alliance was distinct in its membership and mission, funding strategies, and conservation goals but had the shared goal of resource conservation and expanding the WNP. The conservation alliance frequently referred to the 2002 protected areas in Andros as a successful act of environmental protection, stating that the park "paves the way for additional protection in the North and the South" (Bahamas National Trust, 2004). In order to build scientific salience for the expansion plan, the conservation alliance raised funds for the REA of the west side of Andros in 2006.

For the REA, an interdisciplinary team of international researchers gathered evidence of the area's environmental worth such as biological diversity, ecological habitats, and fresh-water reserves. Of particular interest to the conservation agenda were the populations of hawksbill, green, and loggerhead sea turtles, including the only known aggregation of juvenile loggerhead (*Caretta caretta*) turtles in the Wider Caribbean (The Nature Conservancy, 2006a). The research team discovered that the west side may also provide important nursery habitat for various shark species and habitat for the endangered West Indian flamingo (The Nature Conservancy, 2006b). Due to the west side's extreme remoteness and the absence of any infrastructure, the REA research team opted to stay at the one location that could offer accommodation near the research sites, a 5-star hunting lodge located deep in the west-side mangroves, just next to the resident flock of flamingos. The lodge was owned by a well-known wealthy Bahamian family and had been the hunting camp for the Bahamian and British elite for nearly a century before becoming a small remote resort for adventurous recreational fishers. The lodge and its owner became central figures in the dispute over protected area conservation in Andros.

6.2.4 The lodge

Much of the west side of Andros is uninhabited except for a tiny settlement (population 80) far north of the proposed park boundaries and the privately owned commercial hunting and fishing lodge. Accessible only by boat and seaplane, the lodge sits among the broad stand of mangroves and mud, an anomaly in its rugged luxury. The property was acquired by a wealthy Bahamian merchant family in the 1920s as a private hunting and fishing camp and was recently renovated to attract wealthy anglers willing to spend $10,000 (USD) for one week of fishing and hunting off-grid deep in the west-side mangroves. Although the property has been in the family for most of the past century, many Androsian residents do not recognize the owner's claim due to his ancestry as a white settler with colonial ties. The lodge owner was a strong proponent of the original WNP and vocally supported the enlargement plan as evidenced by his volunteering the use of the lodge, boats, and seaplane for the 2006 REA.

After the REA, several outreach campaigns, and extensive discussions among conservationists and scientists, the proposal to expand the park both south- and northward was accepted by the Bahamian government in the fall of 2009. The conflict over the proposal has continued, however. While the Bahamian government officially sanctioned the WNP in May 2012, it is unclear whether the protected area will meet its goals because of ongoing conflict and low levels of compliance for park regulations. Reflecting on the deep-rooted conflict associated with the WNP, this chapter uses the levels of conflict model (Madden and McQuinn, 2014) to examine the claims and counterclaims of rightful ownership—of land, water, and birds—in relation to decision-making processes surrounding protected area conservation in the Bahamas.

6.3 Levels of conflict analysis

6.3.1 The dispute

Although the issues surrounding the WNP began in 2002 with the implementation of the first protected areas in Andros Island and intensified and spread in 2006 with the expansion plans, conflict over the expanding park boundaries is still ongoing. The history of unresolved disputes that is underlying the current dispute is discussed in more detail below.

6.3.2 Underlying conflicts

What began as a series of disputes in 2002 and 2006 are today part of the history of conflict that makes current opposition over protected areas in Andros difficult to resolve. In recent years, there have been both direct acts of resistance and more generalized opposition to protected area conservation in Andros. Signage marking existing protected areas were removed or damaged in some places. There were verbal and public conflicts among conservation managers, resource users, and land owners. Information events were poorly attended and sometimes deteriorated into shouting matches among participants. Tension rose among resource managers and residents, and some conservation partners withdrew their public support for the park. The conservation alliance, which had spanned local and international conservation organizations in support of the WNP expansion, disbanded as some groups withdrew their support. One former alliance leader became a vocal opponent to the expansion, both in the press and during community meetings.

There are several underlying conflicts relating to the WNP expansion. The most obvious—and one that came up frequently during interviews—stems from the implementation process of the initial series of protected areas formed in 2002 (discussed further below). Old wounds formed by conflicts among conservation officials and resource users were made fresh by the renewed discussion over park boundaries and access to resources in 2006 onward.

The 2002 protected area placement process was viewed by many Androsians as a flawed and unjust land grab by a small group of landholders near the proposed parks. Some Androsians interviewed were ambivalent about the established protected areas and argued that the people benefiting most from the parks were the same people who had benefited from the first round of protected areas: private landholders near the parks. Many people took great care to point out, whether on maps or in conversation, the relationship each of these parks had with adjacent properties. During one interview a man said, "I mean, just look at those five parks. Where they at? I mean, I ain't saying nothing, but I'd advise you just to look. You wonder how the science matches up so nice with the property lines, right?" (Interview, Andros resident,

2009). Because of the experience of the 2002 park proposal and implementation, many believed that the local participants most vocally in favor of the parks would benefit directly from the protection of the protected areas through increased land value and more exclusive access to the space. These themes were expressed repeatedly in public meetings and during discussions of park placement. Residents were reluctant to discuss expansion of the WNP, wanting instead to discuss past conservation efforts and park boundaries. Often arguments escalated during these meetings, and little progress or consensus was established concerning the new expansion plan.

Conflict also developed around tension between Androsian residents and the controversial, wealthy, Bahamian businessman who owned a fishing lodge on the west side of Andros. On the vast west side, there was one remote fishing and hunting lodge: a rustic, but high-end fishing resort. The original west-side protected area created in 2002 encircled his lodge property, which was located on a stretch of land long used for resource extraction by Androsians and which was host to a resident flock of West Indian flamingos. Although the land owner had held legal title to the land for nearly a century, his claim was seen as illegitimate by many Androsians who continued to use the area for hunting and fishing and to guide tourists to view the flamingos, against the wishes of the lodge owner. To protect what he saw as his property from what he believed to be unlawful poaching of resources and trespassing, the lodge owner installed fences, hired foreign staff to manage the lodge and police the grounds, and restricted access to the land and sea. The lodge owner's actions were perceived as unjust and therefor resisted. There continued to be deep resentment among many Androsians interviewed about the lodge owner's attempts to claim the land.

To further complicate the issue, the lodge owner continued to support the park expansion plan. Many Androsians were quick to argue that the WNP park boundaries formalized the lodge owner's claim to the land and its resources, including the flamingos. The park expansion would, in practice if not in purpose, enlarge the protected land and sea surrounding the lodge, supporting the view that the park expansion unfairly benefited the lodge owner, and lending fuel to the mounting tensions between him and resident Androsians. People framed their lack of support for the expansion plan through their distrust and dislike for the lodge owner and his policies.

Finally, much of the conflict surrounding the WNP can be traced to the 2006 REA, when a team of international scientists traveled to the west side to study the flora and fauna of the area and document important areas for conservation. The team mapped, measured, and sampled the water, flora, and fauna of the west side of Andros. They found shark and turtle nursery grounds, rare orchids, iguana habitat, and a flock of flamingos numbering from 100 to over 1,000 birds. Because the area can only be reached by a two-hour boat ride or by seaplane, the research team opted to use the controversial hunting and fishing lodge as a launching and resupply station. For many residents, the decision was seen as a partisan move in favor of the wealthy Bahamian's land claim, and against the interests of the Androsian people. People complained that the obvious benefits of the park—in the form of social and economic capital—would only favor wealthy (white) non-Androsians. This simple move to stay at the lodge derailed the park expansion project and contributed to the strong public opposition to the WNP.

The REA sparked the end of the conservation alliance, because of conflicts regarding the allocation of funds and benefits among organizations. Some partners in the original conservation alliance quit the network and withdrew their support. Interviews with residents indicated feelings of suspicion and distrust for the remaining conservation organizations. The very public and contentious dissolution of the conservation alliance framed the public debate over the WNP. In its most polemic positioning, community members were asked to choose between local conservationists—linked to Andros through history, culture, race, and family— and foreign resource managers with few ties to Andros and who were associated with unfair and biased practices regarding land tenure.

6.3.3 Identity-level conflict

While past grievances contributed to the ongoing dispute, identify-level prejudices further complicated the conflict over the WNP. The park became a symbol for conflict over rightful resource ownership and belonging in Andros. The conflict surrounded issues of race, class, wealth, and family ties. Foreign researchers were perceived to be aligned with wealthy white Bahamians by racial affiliation and were flagged as "outsiders." Poor Androsian resource users were seen as "insiders" but lacking understanding of ecological processes.

One of the central sources of identity-level conflict regarding the WNP is the historical context of land tenure and ownership in colonial Bahamas. Land tenure systems in the Bahamas are inextricably tied to slavery, when black bodies were treated as property and the laws of ownership were restricted to white citizens of the Bahamas. Early on, white Bahamian colonists protected their land holdings through legislation, whereas black Bahamians relied on and developed customary tenure to establish commonage and squatter rights (Johnson, 1991; see Comment).

Comment

For an interesting twist, see Chapter 2, in which Booker and Maycock describe how top-down regulations were distrusted in another part of the Bahamas by a group of "outsiders," yachters from other countries who moored in a local harbor.

—The Editors

One form of customary tenure for land held in common is Generation land: land passed down from generation to generation to use by family members. In Andros, Generation land continues to be a highly valued and closely guarded type of property ownership (Bethel, 2000).

Crown land—land held in trust for the Bahamian people by the government—is viewed by many as a form of social and communal security. Androsians claim Crown land to build homes and farm, generate income, and provide security for their families. As conservation agents discussed enclosing a vast expanse of Crown land on the western coast, some Androsians assumed the motive behind the conservation project was to further weaken black Bahamian land holdings and rights. The move to enclose the national park as a protected area was viewed by many as restricting access to common property and a shared good for descendants of enslaved Bahamians: it demonstrated the asymmetries of power among Bahamians and the country's long colonial history of preferential treatment for wealthy white residents. In contrast, resource managers and conservation scientists viewed the ecological landscape as unused, empty, and highly valuable for conservation goals. They assumed the land was valueless because it had not yet been developed. In this way, tenure processes were linked to value-laden ideas about Bahamian identity, belonging, class, and race. These beliefs and values fueled the increased tension surrounding marine conservation in Andros and the wider Bahamas.

One government official in the ministry of the environment articulated his feelings of distrust and injustice surrounding the idea of protected areas in Andros. Although he emphasized the importance of protected areas for environment and society, he also acknowledged the influence of special interests: "But we are also concerned that there are people with selfish interests. It is sometimes in their interest for an area to be protected" (Interview, government official, 2006). The official was referring to the landholders (often wealthy whites) living near the proposed areas and who wanted to control who would have access to the space. His worry was that protected areas would work as buffer zones for private property owners, removing "unwanted" resource users (often poor blacks) under the rubric of conservation.

The government official went on to question the motives of park placement in Andros and linked the concept of protected areas to special interests and discrimination:

> If someone were to tell me they want to put a protected ring around Andros, I have to welcome it, because most of the fish we have is coming from there. . . when someone say, oh, we going to go to Andros, we going to just take this small piece. . . And so it is racism that's built into it. Because it means if you have it protected, it means that the average person in the Bahamas is not going to come to fish in the waters around your property, because they can't take anything. (Interview, government official, 2006)

These remarks underline the conflict surrounding protected areas in general in Andros Island, and more specifically, the WNP expansion plan. For some, the WNP was considered little more than a tool for the wealthy to increase land holdings and restrict access of the "average person in the Bahamas." In this way, protected areas were associated with injustice and exclusion.

Meanwhile, Bahamian conservationists argued in favor of protected areas for the benefit of the native species and natural environment, as well as for the good of the nation as a whole.

The then president of the BNT described protected areas as important to the well-being of the nation as a whole.

> It is so helpful that we have these areas now as our parks to maintain and keep, both for current generations [of Bahamians] and future generations, so that when we look back and see what we have, we know these parks contribute all over the nation because of the diverse marine life that they have. (Smith, 2012)

Conservationists and many locals saw the conflict quite differently. While resource managers promoted the park as a means of protecting resources for future generations, many Androsians perceived WNP as a taking, reducing immediate access to resources necessary for their daily survival for the benefit of an intangible future for only certain well-positioned Bahamians.

A second source of identity-level conflict is the role of racial identity and issues of rightful belonging. The Bahamas has a long history of colonial domination and racial injustice. Racial affiliation is central to national identify, political allegiance, and everyday life. The nation is relatively young, having established independence from Britain in 1973 under a bloodless revolution spearheaded by the People's Liberation Party, a political party promoting black Bahamian leadership and independence from foreign colonial rule. Despite a shift in political representation from white colonial rule to black Bahamian governance, the economic resources remained primarily with a minority of white European descendants. Racial identify informed the dispute of the WNP directly and indirectly. The majority of Androsian residents have African ancestry. However, conservation in the Bahamas was historically led by British colonial subjects, lending to the perception of conservation and resource management as a white project. Although many Bahamian conservation organizations fought against this image by actively hiring black Bahamian administrations and staff, there remained a perception in the Bahamas, and in particularly in Andros, that conservation space was white space. The majority of visiting conservationists and scientists continued to be white, primarily hailing from the United States and Britain, countries directly associated with colonialism in Bahamian history.

One example of this type of identity-level conflict was made visible in the clashes between the west-side lodge owner (a descendant of white colonial landholders) and Androsian residents (primarily descendants of black Africans). The dispute over the WNP expansion was thrown into high relief by the historical significance of race and resource ownership in the Bahamas. The lodge owner held a unique position in the west-side conservation debate as the only deeded landholder on the western shoreline of Andros. The land had been granted to his grandfather during a time when colonial rule was strong and black Bahamians were not afforded the same tenure rights as white citizens. The luxury lodge, accessed only by seaplane or boat, sits among the broad stand of mangroves and mud, a symbol of wealthy white colonial entitlement. The walls of the lodge are lined with images of family, celebrities, and British royalty hunting and fishing in the bush. During an interview, the owner described his labor in building the lodge, his returning visits to the land, first as a boy and later as proprietor, and his ability to "use the west side," to "see what it was truly worth," and to "improve" on its natural beauty.

Meanwhile, many Androsians contested the lodge owner's rightful ownership of the west-side land as a white Bahamian, arguing that only black Bahamians should be eligible for Crown land claims and that he and his family were little more than glorified squatters benefiting from an archaic colonial system steeped in social inequity. Since the 2002 WNP, the land stood as an island of private ownership among protected lands. Access to the surrounding fishing and hunting grounds and flamingo bay was reduced, and the restricted access was sanctioned by the government through the act of enclosure. The lodge owner was a strong supporter of expanding the WNP. He denied that he supported the park to increase his land holdings functionally if not in deed; however, Androsian residents continued to believe his support was motivated by financial concerns rather than conservation.

A third source of identity-level conflict is specific to the island of Andros itself and the assumptions each group made about the other. While some Androsians assumed the conservation effort was led by a social elite group in order to maximize their land claims and benefit financially, conservationists voiced frustration over what they regarded as lack of understanding about conservation goals and a misinformed "lawless" public. During the colonial era, Andros was known as a remote refuge for individuals escaping slavery. Those seeking freedom traveled to the vast and undeveloped lands of Andros to resettle and build a life. Although only 25 nautical miles from the capital of New Providence, Andros offered miles of uncharted and difficult-to-penetrate coppice and marshland. Marronage—the act of escaping and fleeing enslavement—is well documented in Andros (Howard, 2006). The best-known case involved groups of Black Seminole Indians who traveled (some say by dugout canoe) from Florida to the west side of Andros, beginning in 1821. The groups built a small settlement called Red Bays and survived as fishers and spongers. Today, many Red Bays inhabitants can trace their lineage to the original maroons who resisted slavery in the United States and found refuge in the wetlands of western Andros.

On the one hand, Androsians were proud of Andros' reputation as a "lawless" land. For many, it spoke of their independence and historical resistance to the unjust and inhumane system of slavery. Some resource managers and conservationists, however, viewed this legacy of resistance more negatively, instead describing Androsians as "uncooperative" and stubborn. The perception of Andros and its inhabitants as "lawless" was common during interviews with residents as well as with conservation staff and visitors. People frequently referred to the history of piracy and shipwrecking—and the more recent activity of drug trafficking—as examples of how island residents reject formal law. During the time of this research, poaching in protected areas was rampant. In some cases, people were not aware of park boundaries; however, in other cases, poaching may have been a form of resistance to government-led enclosures and property claims. Andros is large and undeveloped, creating challenges to manage and monitor such large tracts of land and sea. There was a general assumption by park staff, scientists, and resource managers that the Androsian population would not comply with regulations regardless of the number of town meetings or outreach campaigns. As a result, low participation rates and lack of consensus was perceived as "the norm" and accepted by conservation agents. Even while attempting to improve stakeholder engagement efforts around the WNP expansion, conservation officials repeated past (failed) attempts to garner support by presenting information about the benefits of the area for conservation, the nation, and global community.

6.4 Conclusions and implications

This chapter uses the levels of conflict model to analyze the underlying and identity-level conflicts involved in the expansion plan for the WNP. Beyond the immediate disputes surrounding the park boundaries and resource tenure, this analysis illuminated some of the historical and social complexities undermining a strong and effective conservation plan for marine resources and the communities that rely on them for their livelihoods. In Andros and elsewhere in the Bahamas, people understand and value protected areas differently. The choice to use the private commercial fishing lodge as lodging for the REA team derailed the conservation project considerably, in part because the nature of the conflict surrounding protected areas in the Bahamas was not fully understood or addressed early in the planning process. The move toward enclosing the west coast of Andros was assumed to be by many Androsians a government-facilitated and racist taking of common property for the benefit of a special-interest group. The 2009 expansion plan emerged from this enduring and complex conflict leading to reduced support or compliance.

The Bahamian social system is hierarchical and multi-tiered. The class structure is organized around race, wealth, education, and historical social networks. As is true of many places around the world, the conservation field is dominated by those with the time and money to devote to environmental causes. Among this group, designating MPAs is considered an appropriate method of resource management and protection: creating boundary lines becomes a way to secure space and protect species and habitat. For many Bahamians who make their living by the sea, these same boundaries fragment and disrupt longstanding and familiar institutional practices, beliefs, understandings, and rules, reallocating access and use rights in specific areas. Land and seas once commonly held are redefined as areas with restricted access. In this way, boundary-making processes that are essential to enclosure conservation become social processes informed and fueled by existing power structures. In the case of the WNP park expansion, the result of underestimating underlying and identity-level conflicts in disputes over resource management was reduced public investment in the management plan, unresolved disputes over the park, and potential lack of compliance with management regulations. At stake is not only a mismatch in marine resource management and ultimately failed policy initiatives, but also a loss, both material and cultural, for the people of Andros. The material consequences consist in the substantial loss of commonage territory used for generations and valuable in daily subsistence. The cultural consequences consist of the loss of Bahamian's sense of agency, security, and well-being. Without the support of resident Androsians, conservation agents risked creating a 1.5 million-acre paper park.

Box 6.1 Encountering flamingos

We left the lodge and headed into a smaller bay. Beneath the boat, the sand turned into a muddy brown and the water, only a few feet deep now, was rusty colored, but still clear. What looked like muddy clumps at the bay's edge took on the pink tones and long necks of flamingos (Figure 6.4). There were hundreds of birds. Last counted in the 1980s, the flock had then numbered 30 birds, but today there were closer to 300. First the adult birds and then the grey juveniles began to move away, before the guide began to back the boat off. We shut off the engine altogether and floated. Mingled with the distant and peculiar clucking of the birds was the sound of the cameras letting off a steady flow of clicks as we tried to document every move. Suddenly, the birds burst into flight and the sky became pink from water's edge skyward. The adults, vibrant pink, necks elongated in flight, elegantly took the lead, while the smaller, grey juveniles flew behind, ungainly, but remarkably beautiful. The birds flew, dipped toward the ground and circled up again and again as we sat in awe at the spectacle. After 20 minutes the birds settled down, first one and then another alighting on the muddy shore to feed. All was calm again and the boat turned back home. For a moment, everyone had lost their heading, their positioning, and their own boundaries of expertise. For a moment, we were united in our awe at being so small, so human, so dwarfed by the vast expanse of the "mud" and the pretty pink birds that dotted its shoreline.

Sarah Wise

Figure 6.4 West Indian flamingo taking flight. Credit: By Joxerra Aihartza (Nire argazki-bilduma) [FAL], via Wikimedia Commons.

Lessons learned

- Historical contexts matter. For example, a background of colonialism can color conflict that may seem from the surface to have little to do with this history.
- Individuals can make a difference. Do not lose sight of the role that individuals can have in driving (or calming) conflict.
- Be aware of the outsider status you might have as a non-local conservation practitioner or researcher, and how your actions might be perceived as such.

—The Editors

References

Agardy, T. (1997). *Marine Protected Areas and Ocean Conservation*. Austin, TX: R. G. Landes Company.

Agardy, T. (2000). Information needs for marine protected areas: scientific and societal. *Bulletin of Marine Science*, **66**(3):875–88.

Bahamas National Trust. (2004). *Central Andros National Parks*. <http://www.bnt.bs/_m1776/Andros>, August 1, 2004.

Bahamas National Trust. (2012). *West Side National Park Draft Management Plan: Review Draft September 2012*. Nassau, Bahamas: Bahamas National Trust.

Bethel, N. (2000). *Navigations: The Fluidity of National Identity in the Postcolonial Bahamas*. PhD thesis. Cambridge: University of Cambridge.

Botsford, L. W., Micheli, F., and Hastings, A. (2003). Principles for the design of marine reserves. *Ecological Applications*, **13**(1):S25–31.

Campbell, L. M., Gray, N. J., Hazen, E. L., and Shackeroff, J. M. (2009). Beyond baselines: rethinking priorities for ocean conservation. *Ecology and Society*, **14**(1):1–12.

Chiappone, M. and Sullivan Sealey, K. M. (2000). Marine reserve design criteria and measures of success: lessons learned from the Exuma Cays Land and Sea Park, Bahamas. *Bulletin of Marine Science*, **66**(3):691–705.

Craton, M. (1962). *A History of the Bahamas*. London: Collins.

Craton, M. and Saunders, G. (2000). *Islanders in the Stream: A History of the Bahamian People: Volume 2: From the Ending of Slavery to the Twenty-first Century*. Atlanta, GA: University of Georgia Press.

Fedler, T. (2010). *The Economic Impact of Flats Fishing in The Bahamas*. <https://igmr.igfa.org/images/uploads/files/Bahamas_Flats_Economic_Impact_Report.pdf>, accessed March 13, 2015.

Halpern, B. and Warner, R. R. (2002). Marine reserves have rapid and lasting effects. *Ecology Letters*, **5**(3):361–6.

Howard, R. (2006). The 'Wild Indians' of Andros Island: Black Seminole legacy in the Bahamas. *Journal of Black Studies*, **37**(2):275–98.

Hulme, D. and Murphree, M. (2001). *African Wildlife & Livelihoods: The Promise and Performance of Community Conservation*. Oxford: James Curry, Ltd.

Interview, Andros resident. (2009). Andros Interview #81. S. Wise.

Interview, government official. (2006). New Providence Interview #33. S. Wise.

IUCN. (2014). *The Andros West Side National Park, Home of the National Bird of the Bahamas*.<http://www.iucn.org/fr/propos/union/secretariat/bureaux/europe/intervenons/europe_outre_mer/?18341/The-Andros-West-Side-National-Park-home-of-the-national-bird-of-the-Bahamas>, accessed January 20, 2015.

Johnson, H. (1991). *The Bahamas in Slavery and Freedom*. Kingston: Ian Randle Publishing.

Madden, F. and McQuinn, B. (2014). Conservation's blind spot: the case for conflict transformation in wildlife conservation. *Biological Conservation*, **178**:97–106.

Mascia, M. (2003). The human dimensions of coral reef marine protected areas: recent social science research and its policy implications. *Conservation Biology*, **17**(2):630–2.

McCay, B. J. (2002). Emergence of institutions for the commons: context, situations, and events. In Ostrom, E., Dietz, T., Dolsak, N., Stern, P. C, S. Stonich, S., and Weber, E. U. (eds), *The Drama of the Commons*. Washington, DC: National Academy Press, pp. 361–402.

Orlove, B. S. and Brush, S. B. (1996). Anthropology and the conservation of biodiversity. *Annual Review of Anthropology*, **25**(1):329–52.

Polunin, N. V. C. and Roberts, C. M. (1993). Greater biomass and value of target coral-reef fishes in two small Caribbean marine reserves. *Marine Ecology-Progress Series*, **100**:167–76.

Roberts, C. M. (1995). Rapid buildup of fish biomass in a caribbean marine reserve. *Conservation Biology*, **9**(4):815–26.

Roberts, C. M., Bohnsack, J. A., Gell, F., Hawkins, J. P., and Goodridge, R. (2001). Effects of marine reserves on adjacent fisheries. *Science*, **294**(5548):1920–3.

Roberts, C. M., Hawkins, J. P., and Gelly, F. R. (2005). The role of marine reserves in achieving sustainable fisheries. *Philosophical Transactions of the Royal Society*, **360**(1435):123–32.

Smith, D. (2012). "Fowl Cays joins list of national parks." *The Tribune* (Nassau, Bahamas), March 13, 2012.

The Nature Conservancy. (2006a). *Andros Island, The Bahamas*. <http://www.nature.org/ourinitiatives/regions/caribbean/bahamas/placesweprotect/bahamas-andros-island.xml>, accessed January 1, 2010.

The Nature Conservancy (2006b). *Rapid Ecological Assessment West Coast of Andros, The Bahamas, June 19th–29th, 2006, Preliminary Findings*. Nassau, Bahamas: The Nature Conservancy.

Walley, C. J. (2004). *Rough Waters: Nature and Development in an East African Marine Park*. Princeton, NJ: Princeton University Press.

West, P. and Brockington, D. (2006). An anthropological perspective on some unexpected consequences of protected areas. *Conservation Biology*, **20**(3):609–16.

West, P., Igoe, J., and Brockington, D. (2006). Parks and peoples: the social impact of protected areas. *Annual Review of Anthropology*, **35**:251–77.

Wilson, J. (2007). Scale and costs of fishery conservation. *International Journal of the Commons*, **1**(1):29–41.

Wise, S. (2014). Learning through experience: non-implementation and the challenges of protected area conservation in The Bahamas. *Marine Policy*, **46**:111–18.

Wood, L., Fish, L., Laughren, J., and Pauly, D. (2008). Assessing progress towards global marine protection targets: shortfalls in information and action. *Oryx*, **42**(3):340–51.

Section 3

NARRATIVES AND HUMAN−WILDLIFE CONFLICT

7

Hawaiian Monk Seals

Labels, Names, and Stories in Conflict

Rachel S. Sprague and Megan M. Draheim

Animals play a vital role in human culture and pervade many aspects of our daily life—both symbolically (through images of animals found in TV and sports, religious iconography, folklore and mythology, etc.) and physically (our food, the pets we share our houses with, the wild animals we see in the landscape, etc.; Herzog and Berghhardt, 1988). In fact, it might be said that mythology, experiences, and folklore play at least as important a role in how we interact with and think about animals as the biology and ecology of the species in question (Kellert et al., 1996). In order to understand our complex relationships with species—including endangered species and those which come into conflict with humans—we need to understand the social constructions that are the lens through which we see them. In this chapter, we discuss the role that stories and narratives play in the way we view animals, using the critically endangered Hawaiian monk seal as a case study to illustrate how humans construct and frame the different levels of conflict around conservation efforts (Chapter 1; Madden and McQuinn, 2014). We will first look at how humans categorize the natural world and how that can play into conservation conflicts, following with a discussion of the levels of conflict model in the specific case of the endangered Hawaiian monk seals. Finally, we will look at the role social constructionism (a social science theory) plays in conflict over conservation in general and, more specifically, with monk seals. Throughout, we will discuss the role that stories and narratives play in monk seal conservation.

Humans tend to categorize both the social and natural world in an effort to make sense of it. For example, by assigning various names to things—whether objects, people, or animals—humans point to how they believe these things should be cataloged (what belongs with what and where). Two categories that often come up in the conservation world are "animals that belong" and "animals that do not belong," or are out of place (Philo and Wilbert, 2000). This "out-of-placeness" does not necessarily reflect the natural world; although it can (such as when referring to native and nonnative species), this rarely tells the whole story. When species that were once native to an area but were later extirpated come back, local residents might see them as out of place and no longer belonging, even if they once did. Oftentimes,

Human–Wildlife Conflict: Complexity in the Marine Environment. Edited by Megan M. Draheim, Francine Madden, Julie-Beth McCarthy, and E. C. M. Parsons © Oxford University Press 2015. Published 2015 by Oxford University Press.

this categorization reflects boundaries between the built environment and "wilderness"—the so-called nature/culture divide (Sabloff, 2001), where wildlife (with a few prized exceptions such as songbirds) is not perceived as belonging to urban and other human-developed areas (Whatmore and Thorne, 1998; Philo and Wilbert, 2000; Jerolmack, 2008; Draheim, 2012). This is not limited to urban environments, however. When wolves migrated back to the French Alps after being extirpated, residents who did not believe that they belonged sought to classify them as "invasive" or "alien" species, seeking to remove governmental protections (Buller, 2008).

The narrative of either belonging or being out of place has serious implications in endangered species conservation, because it both drives and reflects how humans orient themselves to a species and potential conflicts with that species. While conflict over human interests and endangered species is not unique, many histories have come together in the case of Hawaiian monk seals, creating multiple narratives by different human groups of "us vs. them" and ascribing different meanings to monk seals that continue to challenge recovery efforts. Some examples of these histories include the natural history of monk seals, the history of human management of the species, and the sociopolitical history of Native Hawaiians and the U.S. government. This has led to misperceptions about seals, including disagreement among some residents over whether the seals are native to the main Hawaiian Islands in the first place, though in reality they are indeed native to Hawai'i and in fact found nowhere else in the world (Mooallem, 2013). Recently, a 2010 survey found that nearly 40% of people surveyed along Hawai'i's coastlines either did not think monk seals were native to the main Hawaiian Islands or did not know (Sustainable Resources Group International, Inc., 2011). The continuation and spread of a narrative that monk seals do not "belong" can exacerbate animosity and physical violence toward seals; that could threaten this critically endangered species' recovery, as illustrated by the recent suspicious deaths in the main island of four monk seals found with fractured skulls or gunshot wounds, and by evidence of seemingly intentional injuries, such as a young seal found with the prongs of a speargun stuck superficially in her head (Mooallem, 2013). To understand why out-of-place narratives still persist in the face of all evidence to the contrary, we look at the history of monk seals and use the levels of conflict model (Madden and McQuinn, 2014) to examine conservation challenges and controversies.

7.1 Hawaiian monk seal history

The Hawaiian monk seal (*Neomonachus schauinslandi*; Figure 7.1) belongs to one of the oldest groups of seals; of this group, three species have persisted to historical times: the now-extinct Caribbean monk seal, last seen in 1952 (King, 1956); the Mediterranean monk seal, currently numbering only in the hundreds; and the Hawaiian monk seal, currently at a population of about 1,200 individuals and found only in the Hawaiian Archipelago (Figure 7.2). From 150–200 million years ago to about 2.5–3 million years ago, a body of water known as the Central American Seaway existed where Panama is today, connecting the Caribbean Sea to the Pacific Ocean and providing a route by which the monk seals in the Caribbean could enter the Pacific and travel to the Hawaiian Islands (there has been no evidence of monk seals found anywhere else in the Pacific Basin). The Hawaiian monk seal species could be as much as 14 million years old (Repenning et al., 1979), and has almost certainly been in the Hawaiian Islands for at least 3 million years, since the Central American Seaway closed and separated them from their Caribbean cousins. While there is a small amount of archaeological evidence of monk seals in the main Hawaiian Islands prior to European contact (Rosendahl, 1994), Hawaiian monk seals were likely extirpated (or nearly so) from the main islands shortly after the arrival of the first Polynesian colonists around 1100–1290 CE, leaving a remnant population mostly restricted to the uninhabited Northwestern Hawaiian Islands. It is then perhaps not surprising that there is very little evidence of Hawaiian monk seals in Native Hawaiian cultural traditions existing today, though some have been found in traditional songs, oral histories and stories, creation chants and genealogies such as the *Kumulipo* or *Kumu Honua*, or the accepted list of animals that can take the form of an *aumakua*, or ancestor/family protector. Various names that may refer to Hawaiian monk seals include *'īlio-holo-kai* or *'īlio-holo-i-ke-kei* ("dog running in the sea"), *'īlio-holo-i-ka-uaua* ("dog running in the roughness [rough seas]"; Pūkui and Elbert, 1986), *nā mea hulu* ("the furry ones"), and *sila* or *kila* (Hawaiian versions of the English "seal"). The few references and stories that do exist are sometimes subject to disputed interpretations of the translation and are inconsistent and scattered, much like any remaining monk seal population in the main islands would have been after near-extirpation (Watson et al., 2011; Kittinger et al., 2012).

Figure 7.1 Hawaiian monk seal, *Neomonachus schaunislandi*. Tracy Wurth/NOAA NMFS, ESA-MMPA Permit No. 16632-00

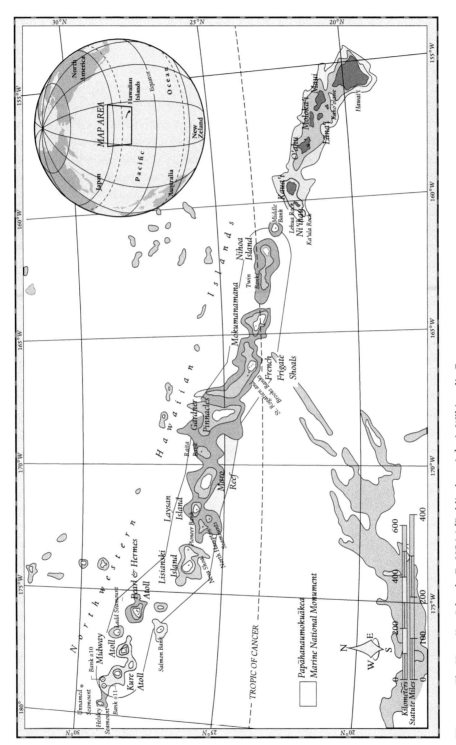

Figure 7.2 The Hawaiian Islands. By NOAA [Public domain], via Wikimedia Commons

The Northwestern Hawaiian Islands monk seal population also sustained heavy losses during the early post-European contact period and was nearly hunted to extinction in the mid-1800s (Rauzon, 2001), though seals were seen again in small numbers in the late 1800s (Bailey, 1952). During the 1800s and 1900s, the main Hawaiian Islands only received visits from transient animals or perhaps had a very small and scattered population, until the species was listed under the U.S. Endangered Species Act in 1976. While the majority of the monk seal population remained in the Northwestern Hawaiian Islands, it was primarily managed there by the U.S. government's National Oceanic and Atmospheric Administration (NOAA) National Marine Fisheries Service (NMFS), which focused on research and interventions to improve survival of the seals (Lowry et al., 2011). As early as 1962, a small, black pup was found abandoned by its mother on Kaua'i (Baker et al., 2011), and the Robinson family on Ni'ihau began seeing seals around their island in the 1970s (Keith Robinson, personal communication). But finally, around 1980, the monk seal population took hold again in the main Hawaiian Islands and began slowly increasing (Kenyon and Rice, 1959; Baker et al., 2011). As seals began to thrive in the main Hawaiian Islands, NMFS maintained a high level of involvement in seal management and continues to be the almost exclusive source of scientific information about monk seal natural history and ecology.

Today, the Hawaiian monk seal is the most endangered marine mammal found entirely within the U.S. jurisdiction, and the population is in crisis: the species is in a decline that has lasted several decades, and only about 1,100 to 1,200 individual monk seals remain (Carretta et al., 2013). Currently, most Hawaiian monk seals (about 900 individuals) reside in an uninhabited MPA consisting of small atolls and submerged banks extending 1,100 miles to the northwest of the main Hawaiian Islands: the Papahānaumokuākea Marine National Monument. However, despite the protected area and current lack of humans, a combination of lingering anthropogenic effects from hundreds of years of human impacts and from natural lows in ecological productivity have pushed the population down a multi-decade decline. While there has been a slight improvement in survival in recent years, approximately 80% of pups born at some locations die before reaching adulthood at about six years of age, leading to an aging and dwindling reproductive population (Baker et al., 2011). In contrast, the Hawaiian monk seal population in the main Hawaiian Islands is much smaller, with only around 150–200 seals, but is growing naturally due to very high juvenile survival to reproductive age. The small population of seals in the main islands is currently one of the most successful in the entire archipelago, but management of seals (and humans) has proved to be extremely complex, and building community support for and ownership of monk seal recovery has been a challenge.

7.2 Levels of conflict

The current conflicts about Hawaiian monk seals are complex and the levels of conflict range from disputes about monk seal management and recovery, to underlying conflict based on a history of distrust and unresolved disputes, to strong identity-based conflict fueled by Hawai'i's history. Their recent recolonization of the main Hawaiian Islands has put Hawaiian monk seals in close contact with humans whose parents and grandparents never saw seals and who do not have a depth of cultural or historical stories of seals. In a presentation at the World Indigenous Network Conference in Darwin, Australia, Dr. Trisha Kēhaulani Watson, a Native Hawaiian activist and consultant, deftly summarized the Hawaiian monk seal conflict: much of the conflict is not about monk seals themselves (though some direct conflict does exist) but over *how* the recovery has taken place (Watson, 2013). The different levels of conflict have resulted in a situation where many people now "see the animal less as an autonomous wild creature than as an extension of the government working to save it" (Mooalem, 2013). These statements clearly illustrate the "levels of conflict" concept introduced by Madden and McQuinn (2014): the conflict is multilayered and is not only about the substance of issues but also about the relationships and process by which that conflict is addressed.

7.2.1 Disputes: resources and regulations

There are legitimate concerns from some stakeholder groups about monk seals in the main Hawaiian Islands; these concerns present a challenge to local groups and communities supporting and promoting monk seal recovery. Some of the conflicts related to Hawaiian monk seals are disputes, or are at the level of straightforward disagreements. For instance, monk seals can directly take bait or catch from fishermen, and there is indirect competition for some of the same fish species (though perceived competition is much greater than the reality); thus, there is the possibility that monk seals could have at least some low level of impact on livelihoods and fishing as an important cultural practice. Some people go further and fear that monk seals are eating all the near-shore fish and think that perhaps seals are the cause of the more than 40-year-long fishery decline in the main islands. However, monk seals' consumption of fish and marine invertebrate biomass is estimated to be about one-third of that landed by near-shore human fisheries (Sprague et al., 2013), and the most economically valuable fish landed by Hawai'i's fisheries are not eaten by monk seals.

Other people may feel positively toward the seals themselves and think that the seals should be protected but are afraid of government regulations that could come along with them, such as beach area closures, critical habitat designation limiting ocean activities, prosecution of people for close proximity or harassment, etc. While some of these fears are unfounded, Hawai'i does have a long history of resource conservation and endangered species protections being connected with limiting access and restricting consumptive resource use (e.g., hunting, fishing, and gathering). Most common species in Hawai'i are invasive—so environmental groups tend to push for their eradication rather than management for hunting or fishing—and most native and endemic species are threatened or endangered, so consumption or harvesting (and in some cases, even approach, disturbance, or physical contact in any way) are prohibited (Lepczyk et al., 2011). Some of Hawai'i's native species are so endangered and sensitive to disturbance or predation that areas around them are made off-limits to any human activity; areas might also be fenced off to exclude nonnative predators and ungulates that damage habitat, limiting human access, hunting, or other resource use, and causing conflict.

On a smaller geographic scale, Hawaiian monk seals can bring a small area of protection with them. Hawaiian monk seals spend much of their time at sea but haul out on beaches to rest and sleep, mostly by themselves or in groups of two or three (Figure 7.3). When they do so, volunteers usually put signs and ropes around the seal in a "seal protection zone" (SPZ; usually a 15–20 ft. radius around the seal) to alert the public so they do not accidentally approach too close and get injured; in this way the volunteers hopefully prevent disturbance of the seal so it may rest (Figure 7.4). SPZs do not close entire beaches and are not legal boundaries; they are only a viewing guideline to help the public adhere to marine endangered species regulations and avoid breaking the law by disturbing the seal. Still, SPZs can be disliked by some beachgoers, who see SPZs as limiting their access to fully enjoy the beach and shoreline.

One recent proposed recovery action sparked a particularly large amount of public debate and intersected with the idea of "belonging" and questions about Hawaiian monk seal origins. In 2010, NMFS proposed a recovery action to help the species that would have involved temporarily bringing up to 20 young seals a year from the Northwestern Hawaiian Islands (where they would only have a 10% chance of survival) to the main Hawaiian Islands for three years (where survival chances are as much as 80%) and then returning them to the northwestern islands before the young seals started breeding (Baker et al., 2013). This "two-stage translocation" proposal was peer reviewed and seen by conservation biologists as one of the most innovative and promising options to date to improve survival of young females in the beleaguered Northwestern Hawaiian Islands. However, it was strongly opposed by members of many local communities, as they felt that it would cause more stress on the marine ecosystem in the

Figure 7.3 Hawaiian monk seal resting on a public beach. Barbara and Robert Billand/NOAA NMFS, ESA-MMPA Permit No. 16632-00

Figure 7.4 Seal protection zone around a Hawaiian monk seal on a public beach. Barbara and Robert Billand/NOAA NMFS, ESA-MMPA Permit No. 16632-00

main islands, and more problems by exacerbating current seal management issues. Similar to the dislike of SPZs and other disputes about monk seal conservation, opposition to two-stage translocation was greatly amplified by underlying and identity-based conflict, and the dispute itself became much, much more than simply the proposed action.

7.2.2 Underlying conflicts

Unfortunately, there is a history of unresolved disputes regarding Hawaiian monk seal conservation that has created underlying conflict. In particular, the underlying conflicts revolve around distrust of the federal government, distrust of restrictions, the issue of moving or translocating seals, and the overarching process of how the public has been engaged (or not) in implementing monk seal recovery. As mentioned in the discussion of disputes, Hawai'i has a history of conservation equating to restrictions or loss of access, unlike some areas of the mainland where conservation of a resource still allows some consumptive use of that resource (e.g., duck stamps for helping to fund wetland restoration and National Wildlife Refuges; Lepczyk et al., 2011).

History, the decision-making processes at play, as well as other nonscientific factors, turned the two-stage translocation proposal into a hot-button issue for many people in communities around Hawai'i, despite its scientific merit and potential benefit for monk seal recovery. NMFS knew that it would be a controversial proposal, but discussion about the proposal quickly became fraught with misinformation and misconceptions, some based on small selective kernels of truth in monk seal recovery history. A comment frequently repeated (with some variations) during public engagement about the translocation proposal was that NMFS had originally introduced Hawaiian monk seals to the main Hawaiian Islands (and that seals did not exist there prior to that), has continued to move seals here (either from the Northwestern Hawaiian Islands or from various other locations, including Alaska and the U.S. mainland, where monk seals do not actually exist) and that NMFS is now trying to convince communities to let them bring even *more* seals down here (Mooalem, 2013). In fact, NMFS did move and release 20 male seals to the main islands in the mid-1990s (Lowry et al., 2011) in an (ultimately successful) attempt to remedy a skew in the sex ratio at Laysan Island in the Northwestern Hawaiian Islands, where some males were ganging up on and severely injuring or killing females and young seals. The translocation was done without much community engagement: public sensitivities and community involvement in compliance with the National Environmental Policy Act, the Endangered Species Act, and the Marine Mammal Protection Act were held to a different standard by both the government and community members nearly 20 years ago. Even though male seals alone cannot affect the population growth rate, the translocation of some seals did happen, and this kernel has served to both complicate current recovery efforts and cast doubt onto assertions by biologists that the population is in fact growing naturally.

In a display of continuing distrust that runs through many issues (not only with monk seals or with NMFS), community members often comment at public meetings that they feel that most government proposals are a "done deal" and that community involvement does not matter or will not be listened to. Along these lines, people have stated at public hearings that they do not want to be asked to simply comment on a proposed action but want their community to be part of the conceptualization of a given project. In Hawai'i (as in other places), efforts are being made to do this, but it can prove difficult to do in reality, because different communities and groups of people often have competing opinions about what is the best way forward. The government attempts to move through regulatory processes, such as the preparation of Environmental Impact Statements with public hearings and comment periods (through the National Environmental Policy Act of 1969), and solicits public comments, while being restricted from conversing with those giving testimony. But after receiving all of the comments supporting or objecting to a certain action, the government agency then decides on a course forward, and inevitably those who objected to that course are left feeling that their comments were dismissed or ignored. Following these federal processes as they are basically outlined in the statues violates community and stakeholder need for genuine inclusion, empowerment, and involvement in decision-making. While the laws need to be followed, there may be times when effort could be made by government agencies to engage communities outside, or in addition to, the official process in order to change the dynamic.

Ideally, traditional Hawaiian practitioners, local communities, and those working to save Hawai'i's native species could be powerful allies. While Hawai'i's pre-European-contact culture may not have lived in perfect harmony with the young islands and delicate species here (e.g., several species of flightless birds went extinct between the arrival of Polynesians and the arrival of Europeans (Milberg and Tyrberg, 1993; Athens et al., 2002)), Hawaiian culture *does* have a strong tradition of sustainable use and deep respect and value for natural resources. It is clear that a different process needs to be undertaken to truly engage and empower communities in endangered species management. But this change can be a challenge when government agencies have legal mandates to implement species recovery actions and may not want, or be able to wait for, the slow process of true community engagement to foster productive participation in conceptualizing conservation management. On the other hand, when taking into account setbacks due to retaliatory actions by those unhappy with a conservation plan they were left out of, the "slow" process might be the quickest way to a resolution after all (see Comment). All it takes is a walk around the Hawai'i Conservation Conference (an annual gathering of Hawai'i's researchers, resource managers, environmental advocates, and native practitioners) to hear a frequently expressed sense of desperation from conservation groups and native species researchers that species are being lost. Hawai'i is called the "extinction capitol of the world," and in their mind there is no time to go about conservation in an inclusive way. At the same time, some cultural practitioners and traditional conservationists/ resource managers at the same conference express frustration that their values and methods of resource protection are not fully acknowledged or accepted by governmental institutions and modern conservationists. "Outside" conservationists may not necessarily trust that traditional Hawaiian resource management will succeed in the current crisis. And the hunters, fishers, and traditional resource managers potentially do not trust that the outside conservationists will not try to limit access and resource use as their primary conservation tool. These underlying conflicts and distrust play through many of the individual efforts to implement management or recovery actions for Hawaiian monk seal and other endangered species.

Comment

For a further discussion of how what at the outset might appear to be a longer process could actually be shorter than "traditional" policy processes, see Jill Lewandowski's work in Chapter 3.

—*The Editors*

7.2.3 Identity

We will talk later in this chapter about how humans construct cultural meanings and understandings of the world around them, and how those constructs can be a major driver of human–wildlife conflict (see Section 7.4). However, humans construct meanings not only of animals but also of other people, and this has direct relevance to many human–wildlife conflict situations, where the social conflict over what to do with wildlife can be much more difficult to sort out than the actual conflict between people and wildlife (Madden, 2004; Harker and Bates, 2007; Draheim, 2012; Madden and McQuinn, 2014). Construction of narratives and assumptions about other groups of people is a concept that fits closely into Madden and

McQuinn's (2014) levels of conflict. Identity-based conflicts are influenced by preconceptions and assumptions about other groups of people (e.g., cultures, ethnicities, and organizations) that are frequently a product of historical conflict. Hawai'i's social and political history has contributed strongly to the identity-based conflicts that are probably the most difficult issues currently driving disagreement over Hawaiian monk seal recovery, as we'll discuss next.

Hawai'i has a rich past, from the development of a complex society and kingdom by Polynesian settlers, and through its more recent history with intricate layers and interactions of culture, religion, politics, economies, and more. Therefore, a simple sketch of some of the major political events after European contact will have to suffice for the purposes of this chapter. When Captain James Cook arrived in Hawai'i in 1778, Hawai'i had local chiefs leading small areas of each island. After European contact, many other interests began coming to Hawai'i, including missionaries, businessmen, immigrant workers, and other governmental representatives. All inhabited islands were united as a monarchy under King Kamehameha the Great after a series of battles ending in 1795 (and cession of Kaua'i in 1810). The House of Kamehameha ruled the Kingdom of Hawai'i until 1872. As the missionaries started converting Hawaiians to Christianity in earnest in the early 1800s, traditional Hawaiian practices and knowledge were lost or suppressed (Trask, 1999) via, for example, bans on hula and speaking or teaching the Hawaiian language in public schools. These effects were long-lasting: for example, the language ban remained on the books until the 1970s. In the mid-1800s, foreign business interests orchestrated a redistribution of land from a state of no-ownership to a system where the king and *ali'i* (chiefs) could divide land among themselves and ultimately sell that land (known as "The Great Māhele").

The 1887 Constitution of the Kingdom of Hawai'i, known as the "Bayonet Constitution" and signed by King Kalākaua under threat of violence, stripped the king of authority and disenfranchised Hawaiians and immigrant laborers to favor the wealthier white community. Finally, in 1893, Queen Lili'uokalani was overthrown, again under threat of violence, by American business and political interests who were threatened by the queen's attempts to restore power to the monarchy and reduce American and European voting influence. The sugar trade had boomed in Hawai'i during the U.S. Civil War, when production by southern states waned. So annexation also provided economic benefits to U.S. businessmen in Hawai'i by allowing them to stay competitive in the sugar market through more favorable tariffs. Faced with potential bloodshed, the queen yielded the throne—she believed only temporarily—with the following statement:

> Now to avoid any collision of armed forces and perhaps the loss of life, I do, under this protest, and impelled by said force, yield my authority until such time as the Government of the United States shall upon facts being presented to it undo the action of its representatives and reinstate me in the authority which I claim as the constitutional sovereign of the Hawaiian Islands. (Kuykendall, 1967, p. 603)

It should be clear from even this brief history why Hawai'i is starting off from a place of strong identity-based conflicts—for example, between Native Hawaiians and non-Hawaiians, between island residents and those who have come from elsewhere (particularly the U.S. mainland), and between island residents and the U.S. federal government. In a modern conservation context, many of these identity-based conflicts figure heavily. For Hawaiian monk seals, some communities hold a strong association between monk seals and the federal government and mainland conservation organizations. This association, combined with the comparative lack of seal presence in traditional Hawaiian culture, may have reinforced the idea that persists today that monk seals are not a native and endemic species but something that has come to Hawai'i or that has been brought and introduced to the islands by the federal government (i.e., NMFS). Other than the government agencies in Hawai'i, the strongest push for Hawaiian monk seal recovery continues to come from forces outside Hawai'i. For instance, some large conservation groups occasionally push NMFS for more monk seal protection and sometimes disagree with NMFS on recovery priorities. These groups are often based on the mainland and have been seen by Hawai'i residents as bringing outside (and unwelcome) pressure and influence to local issues, sometimes even initiating processes or issues that NMFS and local conservation groups would prefer to avoid. So monk seals ultimately end up being seen by some as being championed by groups of people that are themselves considered "out of place" or not belonging.

Early government efforts toward monk seal conservation inadvertently played into these identity conflicts and have amplified current problems. For example, as discussed earlier in the chapter, early seal translocations without much stakeholder involvement created underlying conflict during a more recent proposal by NMFS for a different kind of translocation (i.e., the temporary movement of young seals to the main Hawaiian Islands and back to the Northwestern Hawaiian Islands as juveniles to increase survival). Obviously, the more recent proposal to bring seals to the main islands, especially by the federal government, then played into the "out-of-place" narratives and identity conflicts.

As another example of monk seal conservation efforts playing directly into existing identity conflicts, when the monk seal population started increasing again in the main Hawaiian Islands, NMFS began recruiting community volunteers. These volunteers help respond to seals on the beach by posting signs and ropes to identify the seals for public safety reasons (many people don't notice lone seals on the beach or aren't aware of the dangers involved in approaching a wild animal) and to help prevent disturbance of the seals. However, most volunteers are retired Caucasians from the mainland who now live in Hawai'i and have time to go out to beaches and watch over the seals. While volunteers are technically "local" Hawai'i residents, there is a long-standing and understood difference between locals who were born here (and are more likely of Asian or Pacific Islander decent) and those who are of European descent who moved to Hawai'i. Public experiences with volunteers can often be extremely positive and involve outreach and the gain of a new monk seal advocate/supporter, but territorial disputes can also occur between volunteers and locals disagreeing over monk seal protection versus perceived beach ownership and access rights. Volunteers donate their time because they are extremely passionate about Hawaiian monk seal protection, not necessarily about people. That passion can sometimes lead to erring on the side of protecting the seal; for instance, using ropes and signs to set aside much more space around the seal than NMFS biologists say is required or necessary to prevent disturbance. Between mainland-based conservation groups, a federal agency as the central push for monk seal recovery and the sole source for monk seal information, and a volunteer community made up of many people who have moved from the mainland, monk seal protection is further associated with outsiders.

7.3 Hawaiian monk seals in the context of constructionism and framing

We will now turn our attention to a social science theory that provides a basis for understanding some of the ways we give meaning to animals; these meanings help in part to construct the conflict surrounding animals. Social constructionism has been both useful and controversial in the conservation biology world (Evernden, 1992; Goedeke and Herda-Rapp, 2005; Hannigan, 2006). A constructionist approach stipulates that humans create cultural meanings and understandings of the world: places, objects, animals, events, other people, and so on. These cultural meanings are not necessarily tightly linked to scientifically observed reality, and this is perhaps where conservation biologists historically expressed skepticism of the theory (Goedeke and Herda-Rapp, 2005; Hannigan, 2006). Some social scientists and philosophers go so far as to question whether or not there is a material world. However, social scientists concerned with conservation biology and other environmental issues by and large take a much more moderate approach by acknowledging that there is a "real world," above and beyond social and cultural constructions, but that it is necessary to understand the multiple layers of social meaning attached to objects, places, and events of interest if environmental conflict is going to be resolved (for a complete discussion of this controversy, see Hannigan, 2006). Dryzek (2005, p. 12) sums it up nicely: "Just because something is socially interpreted does not mean it is unreal. Pollution does cause illness, species do become extinct, ecosystems cannot absorb stress indefinitely, tropical forests are disappearing. But people can make very different things of these phenomena and—especially—their interconnections, providing grist for political dispute." Scarce (2000, p. 8) adds: "It cannot be emphasized too strongly that *meanings are what is being constructed*, not 'things-in-themselves'; or pure material reality untouched by human artifice" (italics original). In other words, the number of Hawaiian monk seals left is not necessarily in dispute, nor is whether the populations across the main and Northwestern Hawaiian Islands are growing or shrinking. What is in dispute is whether or not those populations are worth protecting, and at what cost.

Hawaiian monk seals are not the only species viewed through the lenses of human values and identities (which can be very different even between groups of humans located in the same geographic area). Additionally, the same behaviors of a particular species can be given different meanings by different groups of people (Goedeke, 2005; Draheim, 2012). For example, Draheim (2012) found that some residents of suburban Denver viewed coyotes as being vicious, wicked, and criminal when coyotes attacked pets. Others felt that it was a natural process for coyotes to kill prey-sized animals when given the chance, whether or not humans had emotional attachment to them. Not surprisingly, the two sides often came into conflict with each other. In a similar way, some might see seals taking fish from nets or lines as being a naturally opportunistic action for a wild animal to take that has the same right to fish in the ocean as humans, while others might see this as a criminal event and label it as such—the seal is "stealing" someone else's hard-earned fish.

The example of the seal taking or stealing fish would be described, or "framed," very differently depending on who is discussing the situation. Framing an issue occurs when an individual or group wants to define a problem in a particular way and so adopts particular language to describe the issue as they see it (Goffman, 1986). How different social groups identify (or not) with wildlife also affects how animals are labeled and determines the narratives that are created about them. At times, both sides of a conservation issue will use similar frames to plead their case. Campion-Vincent (2005) showed that all sides of a dispute over wolves in the French Alps claimed to be most concerned with protecting nature; however, what constituted protecting nature, and indeed what constituted nature, was understood differently by those who were in favor of wolf populations and those who were against. Schreiber et al. (2003) also demonstrated how opposing sides of salmon aquaculture in British Columbia, Canada, used similar persuasive frames (those that they felt would resonate with their targeted audiences) to make their cases; the specifics of their arguments differed, but the major framing mechanisms were the same for those who supported aquaculture and those who were against it.

Advocates and opponents of Hawaiian monk seals also sometimes use persuasive frames to resonate with their targeted audience. This can result in conflicts that are not necessarily being debated as two opposing sides of an issue but rather as lines of framing that do not necessarily intersect. Hawaiian monk seal conservationists and advocates generally try to use the facts of monk seal ecology to argue that negative impacts of the seals are not as significant as may be perceived. However, the suspicion and dislike of the federal government has given a convenient excuse for some to voice distrust and dismissal of the science that describes Hawaiian monk seal biology and natural history, conducted almost entirely by NMFS. An article published on June 9, 2013, in Kaua'i's newspaper *The Garden Island* notes that monk seal opponents "say they are being fed propaganda, often by a group of people who, like the seals, aren't Hawaiian."

Hawai'i has a rich tradition of cultural knowledge being passed on orally, as well as a tradition of respect for hearing other people's thoughts and opinions, or *mana'o* (Pūkui and Elbert, 1986). In the absence of commonly accepted traditional stories and traditional cultural knowledge about Hawaiian monk seals, these traditions can combine to elevate opinions and perceptions by contemporary Hawaiians and discount evidence that has been derived from a nontraditional cultural source (i.e., Western science; however, see Box 7.1 for different *mana'o* from a well-known Hawaiian conservationist). A local Kaua'i fisherman who does not believe monk seals are native to Hawai'i was quoted by *The Garden Island* in the same 2013 article as saying, "All these people, who come from somewhere else, they all have their opinions. That's all it is, just an opinion." This strategy does not directly counter the scientific information but uses a tradition that people can culturally identify with to dispute the validity of the science. This can become a popular line of argument, as it is attractive to those who may be somewhat ambivalent toward monk seals but feel that their identity, culture, and way of life are threatened by the federal government and other outsiders. As we will see, it is not uncommon for people to link concerns over animals to greater societal concerns, which is also at issue with debates over monk seal recovery efforts.

Box 7.1 Sharing of one person's *mana'o* about Hawaiian monk seal conservation

Though I have found no reference to them in *mo'olelo* [oral traditions] or *Kumulipo* [one of the Hawaiian creation chants listing native plants, animals, and a genealogy of Hawaiian royalty], my appreciation for *nā mea Hawai'i* is not limited to those reference points. *Kumulipo* is not an exhaustive listing, and *mo'olelo* are bound in time and space and experience and perhaps I will yet learn of a traditional reference. In our time and through our experiences in *Hawai'i nei*, new poetry may be chanted, history may be recounted from a yet unspoken perspective, surely they may be referenced there. There is sufficient biological and experiential diversity under the heading of "Native to Hawai'i," that the Hawaiian monk seal cannot be anything but a valued component of it. There is an alarming diminishment of native biota and spontaneous interaction with it, that the Hawaiian monk seal cannot but be valued as a remainder of the remaining biota, even as it adjusts to the changes of the environment and man. I am native born, I am Hawai'i, no less so the seal.

Hannah Kihalani Springer, Kama'aina of Ka'ūpūlehu, Kona 'Akau (Hawai'i Island)
*Brackets added for clarification.

7.4 Human–wildlife conflict, narratives, and current social concerns

Humans have a long history of linking wildlife to other contemporary social concerns and the values or issues of the day. For example, Jerolmack (2008) looked at the changing way pigeons were described in New York City over almost a century. The imagery of "rats with wings" only became the dominant discourse in the 1930s, when pigeons were used as a metaphor for other perceived social "problems" of the day, including homosexuals, alcoholics, and the homeless. Another bird, the English sparrow, was likewise linked to a social "problem" in the late 1800s—that of immigration and immigrants (Fine and Christoforides, 1991). The authors of this study describe the use of metaphorical linkage by those who participated in the so-called "English sparrow war," where nature enthusiasts and the government alike attempted to eradicate English sparrows from the American landscape, often comparing them to unwelcome immigrants. Metaphorical linkage occurs when people use "metaphor to connect an emerging social problem with another well-recognized problem. While on occasion the two problems are structurally quite similar . . . linkage can occur when the problems are ostensibly very dissimilar" (Fine and Christoforides, 1991, p. 376). Linking a well-understood issue to a new social issue is an effective way to give the newer concern legitimacy. This does not mean that people deliberately frame issues using metaphorical linkage; rather, some people most easily understand the issue in those terms (Fine and Christoforides, 1991). The English sparrow war often directed attention to the English sparrow's status as a nonnative bird in America. Many wrote at the time that sparrows drove out native birds which belonged to the land and held positive, admirable traits. The sparrows, on the other hand, were perceived as being out of place and as having undesirable characteristics, which linked them to negative stereotypes of recent immigrants.

Today in Hawai'i, the socioeconomic challenges include underlying and identity-based conflicts primarily revolving around the struggle to maintain traditional Hawaiian cultural practices, and conflict between local/native versus "other." It is against this social backdrop that the recent narratives about Hawaiian monk seals have arisen. While many other marine species (including sharks and sea turtles) have established cultural and religious significance to Native Hawaiians (Pūkui et al., 1983) and consequently are afforded some "protection" from more recent reframing, monk seals rarely fall into this category (Watson et al., 2011; Kittinger et al., 2012), in part due to a lack of a strong representation of monk seals in pre-European contact Hawaiian traditional chants (*oli*) and oral traditions (*mo'olelo*). Between the linkage with "out-of-place" government agencies and conservation organizations, and little representation in traditional Hawaiian culture, monk seals have been vulnerable to narratives that frame them as a nonnative, or even invasive, species that directly and indirectly threatens other native species that do have cultural significance (e.g., some fish that are prey to monk seals, sharks that are being removed because of predation on seal pups, etc.). People want to identify with and protect native animals that exemplify Hawai'i's unique cultural nature, not one that calls to mind outside influence and anger for all that has been lost.

As society changes, issues of identity often become particularly important in conservation conflicts (Scarce, 2005; Buller, 2008; Wieczorek Hudenko, et al., 2008). Other long-term social concerns can also frame wildlife in a particular way. For instance, Clark et al. (1996) use a political and cultural standpoint to describe the importance of private property rights to predator conservation in the American West. Social change in Hawai'i is particularly fraught with layers of meaning, given the history of the islands. Some social scientists have observed that traditional culture has been portrayed—at least in the large local tourism sector—in "Disney-like versions of past Polynesian cultures" (Rosenbaum and Wong, 2008, p. 174). The so-called "Bali syndrome" (Minca, 2000) puts forward that tourists do not necessarily want to learn about real life in vacation destinations but rather want some color to their vacation settings, even if that color is not historically or contemporarily accurate. The visitor industry is the largest contributor to Hawai'i's economy, so these not-necessarily-sanctioned representations of Hawaiian culture are not geographically isolated but can be seen throughout communities across the islands. This has created somewhat of a backlash in the Hawaiian community against the commercialization of traditional cultural values and practices (Trask, 1999; Council for Native Hawaiian Advancement, 2011). Ownership of cultural traditions then becomes increasingly important in Native Hawaiian communities and contributes to resistance toward information or interpretations of traditional cultural knowledge coming from Western-educated researchers. Given Hawai'i's colonial history, some communities likely feel an acute loss of cultural identity, and as we have seen, wildlife is often used as a metaphor for such societal issues. The tourism industry brings in a large proportion of the state's revenue. But local residents sometimes express mixed feelings toward tourism, because while visitors bring money to the state's economy, many actions are taken for their sake, not for the sake of the state's residents. Hawaiian monk seal conservation can be seen in this way; some feel that the seals contribute to ecotourism, prompting the government to protect them. At the same time, the seals are seen as causing difficulties for local fishermen, and that causes concern that traditional cultural practices may be negatively impacted. In this way, larger societal issues and concerns over monk seal recovery become intertwined; it is necessary to consider both before a full understanding can be had (Box 7.2).

Box 7.2 Using subtle changes in interactions to change the conflict dynamic

Even simple interactions on the beach between volunteers and beachgoers can sometimes exacerbate identity conflicts and leave local residents and Native Hawaiians feeling marginalized when conversation is focused on the seals. When asked, some local people say that the primary reason they do not like monk seals is because they feel that the seals are more valued by the government, conservation organizations, and society-at-large than are the local people and Native Hawaiians, not because of the reasons the National Marine Fisheries Service or the conservation community might think (e.g., they think seals are not native or that the seals eat too many fish). For people who are already feeling disenfranchised, marginalized, and disrespected, small subtleties in casual interactions can take on greater meanings that reinforce or reduce existing conflicts. In light of this, consider two monk seal volunteer interactions with the public. In the first, a volunteer approaches someone on the beach and politely says, "Can you please make sure that you stay away from the endangered seal sleeping over there? It is important to let them rest." In the second, the volunteer changes the overt focus of the interaction away from the seal's protection: "Hi, my name is Jane. Do you come to this beach often? Have you seen a Hawaiian monk seal before? I want to make sure you and your family are safe, so be sure you give the seal some space while you're out here enjoying the beach today." Acknowledging the person and their place on the beach might be subtly more respectful and meet people's need to feel acknowledged and valued.

7.5 Conclusion

The case of Hawaiian monk seal conservation in the main Hawaiian Islands is an example of how social constructions can impact conservation efforts. Using the levels of conflict model, one can see how social constructs of the species, stakeholders, and history have affected the course of monk seal conservation over the last several decades. In order to move forward with recovery, it is essential to understand how monk seals and the social conflict around them are framed and constructed by different stakeholders. If Hawaiian monk seals are to maintain their foothold in the main Hawaiian Islands, their protection and stewardship must become something that is not driven by laws and the government but by efforts both to transform (in adherence to current laws) the underlying and identity conflicts that impede current decision-making and to empower and cultivate community involvement and ownership over conservation efforts. Recovery efforts should work to improve inclusive, positive associations between Hawaiian monk seals and traditional conservation values of respect and protection and take into account the narratives and social constructions stakeholders use when thinking about these animals.

> **Lessons learned**
>
> - People often consider wildlife as either "belonging" or "not belonging" to a particular area in a way that often has little to do with ecological reality; animals that are seen as not belonging are sometimes not valued or even actively disliked.
> - At times, conflict over wildlife is described in terms that relate to a well-established social issue. Looking for these connections can help illuminate both the depth of the conflict and how the stakeholders perceive themselves to be situated with it.
> - Wildlife that is seen as a threat to culturally important species might be poorly tolerated, even if it is endangered.
>
> *—The Editors*

References

Athens, J. S., Tuggle, H. D., Ward, J. V., and Welch, D. J. (2002). Avifaunal extinctions, vegetation change, and Polynesian impacts in prehistoric Hawaii. *Archaeology in Oceania*, **37**(2):57–78.

Bailey, A. M. (1952). The Hawaiian monk seal. *Museum Pictorial, Denver Museum of Natural History*, **7**:1–32.

Baker, J. D., Harting, A. L., and Littnan, C. L. (2013). A two-stage translocation strategy for improving juvenile survival of Hawaiian monk seals. *Endangered Species Research*, **21**(1):33–44.

Baker, J. D., Harting, A. L., Wurth, T. A., and Johanos, T. C. (2011). Dramatic shifts in Hawaiian monk seal distribution predicted from divergent regional trends. *Marine Mammal Science*, **27**(1):78–93.

Buller, H. (2008). Safe from the wolf: biosecurity, biodiversity, and competing philosophies of nature. *Environment and Planning A*, **40**(7):1583–97.

Campion-Vincent, V. (2005). The restoration of wolves in France: story, conflicts, and uses of rumor. In Herda-Rapp, A. and Goedeke, T. (eds), *Mad About Wildlife: Looking at Social Conflict over Wildlife*. Boston: Brill Academic Publishers, pp. 99–122.

Carretta, J. V., Oleson, E., Weller, D. W., et al. (2013). *U.S. Pacific Marine Mammal Stock Assessments: 2012*. U.S. Department of Commerce, NOAA Technical Memorandum, NMFS-SWFSC-504.

Clark, T. W., Curlee, A. P., and Reading, R. P. (1996). Crafting effective solutions to the large carnivore conservation problem. *Conservation Biology*, **10**(4):940–8.

Council for Native Hawaiian Advancement. (2011). *Cultural Strength: The Foundation for Native Economies Observations from the Native Hawaiian Experience*. <http://www.hawaiiancouncil.org/docs/culturefoundation111604.pdf>, accessed March 14, 2015.

Draheim, M. M. (2012). *Social Conflict and Human–Coyote Interactions in Suburban Denver*. Ph.D. thesis. Fairfax, VA: George Mason University.

Dryzek, J. S. (2005). *The Politics of the Earth: Environmental Discourses*. Oxford: Oxford University Press.

Evernden, N. (1992). *The Social Creation of Nature*. Baltimore, MD: Johns Hopkins University Press.

Fine, G. A. and Christoforides, L. (1991). Dirty birds, filthy immigrants, and the English Sparrow War: metaphorical linkage in constructing social problems. *Symbolic Interaction*, **14**(4):375–93.

Goedeke, T. L. (2005). Devils, angels or animals: the social construction of otters in conflict over management. In Herda-Rapp, A. and Goedeke, T. (eds), *Mad About Wildlife: Looking at Social Conflict over Wildlife*. Boston: Brill Academic Publishers, pp. 25–50.

Goedeke, T. L. and Herda-Rapp, A. (2005). Introduction. In Herda-Rapp, A. and Goedeke, T. (eds), *Mad About Wildlife: Looking at Social Conflict over Wildlife*. Boston: Brill Academic Publishers, pp. 1–24.

Goffman, E. (1986). *Frame Analysis*. Boston: Northeastern University Press.

Hannigan, J. (2006). *Environmental Sociology*, 2nd edn. New York: Routledge.

Harker, D. and Bates, D. C. (2007). The black bear hunt in New Jersey: a constructionist analysis of an intractable conflict. *Society and Animals*, **15**(4):329–52.

Herzog, H. A. and Burghardt, G. M. (1988). Attitudes toward animals: origins and diversity. In Rowan, A. N. (ed.), *Animals and People Sharing the World*. Hanover, NH: University Press of New England, pp. 75–94.

Jerolmack, C. (2008). How pigeons became rats: the cultural-spatial logic of problem animals. *Social Problems*, **55**(1):72–94.

Kellert, S. R., Black, M., Rush, C. R., and Bath, A. J. (1996). Human culture and large carnivore conservation in North America. *Conservation Biology*, **10**(4):977–90.

Kenyon, K. W. and Rice, D. W. (1959). Life history of the Hawaiian monk seal. *Pacific Science*, **13**:215–52.

King, J. E. (1956). The monk seals. *Bulletin of The British Museum (Natural History)*, **3**(5):220.

Kittinger, J. N., Bambico, T. M., Watson, T. K., and Glazier, E. W. (2012). Sociocultural significance of the endangered Hawaiian monk seal and the human dimensions of conservation planning. *Endangered Species Research*, **17**(2):139–56.

Kuykendal, R. S. (1967). *The Hawaiian Kingdom 1874–1893: The Kalakaua Dynasty*. Honolulu: University of Hawaii Press.

Lepczyk, C. A., Johnson, E. D., and Hess, S. (2011). Is the model a misfit in Hawaii? The North American model in our most recent state. *The Wildlife Professional*, **5**(3):64–7.

Lowry Lloyd F., Laist, D. W., Gilmartin, W. G., and Antonelis, G. A. (2011). Recovery of the Hawaiian monk seal (*Monachus schauinslandi*): a review of conservation efforts, 1972 to 2010, and thoughts for the future. *Aquatic Mammals*, **37**(3):397–419.

Madden, F. (2004). Creating coexistence between humans and wildlife: global perspectives on local efforts to address human–wildlife conflict. *Human Dimensions of Wildlife*, **9**(4):247–57.

Madden, F. and McQuinn, B. (2014). Conservation's blind spot: the case for conflict transformation in wildlife conservation. *Biological Conservation*, **178**:97–106.

Milberg, P. and Tyrberg, T. (1993). Naïve birds and noble savages – a review of man-caused prehistoric extinctions of island birds. *Ecography*, **16**(3):229–50.

Minca, C. (2000). 'The Bali Syndrome': the explosion and implosion of 'exotic' tourist spaces. *Tourism Geographies: An International Journal of Tourism Space, Place and Environment*, **2**(4):389–403.

Mooallem, J. (2013). "Who would kill a monk seal?" *New York Times Magazine*, May 8, 2013, <http://www.nytimes.com/2013/05/12/magazine/who-would-kill-a-monk-seal.html?smid=pl-share>, accessed May 8, 2013.

Philo, C. and Wilbert, C. (2000). Animal spaces, beastly places: an introduction. In Philo, C. and Wilbert, C. (eds), *Animal Spaces, Beastly Places: New Geographies of Human–Animal Relations*. New York: Routledge, pp. 1–14.

Pūkui, M. K. and Elbert, S. H. (1986). *Hawaiian–English Dictionary*. Honolulu: University of Hawaii Press.

Pūkui, M. K., Haertig, E. W., and Lee, C. A. (1983). *Nana I Ke Kumu (Look to the Source)*, Vol. 1. Honolulu: Hui Hanai.

Rauzon, M. J. (2001). *Isles of Refuge: Wildlife and History of the Northwestern Hawaiian Islands*. Honolulu, HI: University of Hawaii Press.

Repenning, C. A., Ray, C. E., and Grigorescu, D. (1979). Pinniped biogeography. In Gray, J. and Boucet, A. J. (eds), *Historical Biogeography, Plate Tectonics and the Changing Environment*. Corvallis, OR: Oregon State University Press, pp. 357–69.

Rosenbaum, M. S. and Wong, I. A. (2008). When tourists desire an artificial culture: the Bali Syndrome in Hawaii. In Woodside, A. and Martin, D. (eds), *Tourism Management: Analysis, Behavior and Strategy*. Cambridge, MA: CABI, pp. 174–84.

Rosendahl, P. H. (1994). Aboriginal Hawaiian structural remains and settlement patterns in the upland archaeological zone at Lapakahi, Island of Hawaii. *Journal of Hawaiian Archaeology*, (3):15–70.

Sabloff, A. (2001). *Reordering the Natural World: Humans and Animals in the City*. Toronto: University of Toronto Press.

Scarce, R. (2000). *Fishy Business: Salmon, Biology, and the Social Construction of Nature*. Philadelphia, PA: Temple University Press.

Scarce, R. (2005). More than mere wolves at the door: reconstructing community amidst a wildlife controversy. In Herda-Rapp, A. and Goedeke, T. (eds), *Mad About Wildlife: Looking at Social Conflict Over Wildlife*. Boston: Brill Academic Publishers, pp. 123–46.

Schreiber, D., Matthews, R., and Elliott, B. (2003). The framing of farmed fish: product, efficiency, and technology. *Canadian Journal of Sociology*, **28**(2):153–69.

Sprague, R. S., Littnan, C. L., and Walters, J. S. (2013). *Estimation of Hawaiian Monk Seal Consumption in Relation to Ecosystem Biomass and Overlap with Fisheries in the Main Hawaiian Islands*. U.S. Department of Commerce, NOAA Technical Memorandum, NOAA-TM-NMFS-PIFSC-37.

Sustainable Resources Group International, Inc. (2011). *Public Perception and Attitudes About the Hawaiian Monk Seal: Survey Results Report (Final), April 2011*. Prepared for the Protected Resources Division, NOAA Fisheries Pacific Islands Regional Office. <http://www.fpir.noaa.gov/Library/PRD/Hawaiian%20monk%20seal/MonkSeal_SurveyResults_Final.pdf>, accessed March 14, 2015.

Trask, H. K. (1999). *From a Native Daughter: Colonialism and Sovereignty in Hawaii*. Honolulu: University of Hawaii Press.

Watson, T. K. (2013). *Aloha Aina: A Framework for Biocultural Resource Management in Hawaii's Anthropogenic Ecosystems*. Presentation at the World Indigenous Network Conference, Darwin, Australia, May 26–31.

Watson, T. K., Kittinger, J. N., Walters, J. S., and Schofield, T. D. (2011). Culture, conservation, and conflict: assessing the human dimensions of Hawaiian monk seal recovery. *Aquatic Mammals*, **37**(3):386–96.

Whatmore, S. and Thorne, L. (1998). Wild(er)ness: reconfiguring the geographies of wildlife. *Transactions of the Institute of British Geographers*, **23**(4):435–54.

Wieczorek Hudenko, H., Decker, D. J., and Siemer, W. F. (2008). *Stakeholder Insights into the Human-Coyote Interface in Westchester County, NY*. HDRU Series Publication 08-1. Ithaca, NY: Human Dimensions Research Unit, Department of Natural Resources, Cornell University.

8

Flipper Fallout

Dolphins as Cultural Workers, and the Human Conflicts that Ensue

Carlie Wiener

8.1 Introduction

Human desire to interact with dolphins has existed for centuries; however, it is only in the past 40 years that the reputation of dolphins through popular culture has perpetuated a growing dolphin swim tourism industry. Attitudes toward marine mammals began to shift in urban Western society in the late twentieth century as knowledge of personal relationships with dolphins began to surface (Silva, 2013). Public persona of the dolphin as a fused portrayal between an exotic, wild creature and intelligent human-like species has played an important role in driving dolphin swim tourism. The recent appeal of "swimming with dolphins" has created globally distributed hotbeds of tourism activity, transforming marine environments into popular destinations (Wiener, 2013). Additionally, the use of the dolphin figure in new-age mysticism has further increased the dolphin's popularity, linking dolphin swim experiences to human–animal telepathy and healing.

Dolphin–human connections are central to present-day dolphin tourism, but interspecies relationships are not new, and there are many examples throughout history. The dolphin was a popular figure in ancient Greek and Roman society and was depicted as both wise and sacred (Sacks, 1995). Dolphin legends are found globally and portray these marine mammals as mythic figures, friends to humans, hunted animals, and connectors to the natural world. Early literary references to dolphins are common in fables such as Aesop's *The Monkey and the Dolphin* (Montagu and Lilly, 1963). Dolphins have reemerged in the twenty-first century as a central figure in popular culture, taking on several personas that reinforce human desire for interaction. The portrayal of dolphins as helpers, jesters, happy beings, and boundary crossers has perpetuated human engagement, some of which has led to people believing that dolphins have medicinal powers and telepathic abilities.

The morality of wild dolphin swims has been the impetus for substantial debate that advances the question of whether there are negative impacts of this activity on dolphins (Constantine, 2001; Bejder et al., 2006; Courbis and Timmel, 2009; Christiansen et al., 2010;

Human–Wildlife Conflict: Complexity in the Marine Environment. Edited by Megan M. Draheim, Francine Madden, Julie-Beth McCarthy, and E. C. M. Parsons © Oxford University Press 2015. Published 2015 by Oxford University Press.

Donaldson et al., 2012). Of primary concern is the possibility that continued expansion of swim tourism may bring harm to wild dolphins by directly modifying their behavior (Cloke and Perkins, 2005). Conflicting values between those who seek out dolphins to participate in swimming activities and those who work on regulating human interactions with dolphins has led to community conflicts in many dolphin swim sites. These conflicts are deeper than just the values surrounding swimming with dolphins and play off long-time community relationships. One notable location where this conflict has been ongoing for over 40 years is the Hawaiian Islands. Using the levels of conflict model, the long-term struggle between dolphin swim communities in the Hawaiian Islands will be explored in this chapter (Chapter 1; Madden and McQuinn, 2014).

8.2 A case study of Hawaiian spinner dolphin swim tourism

In Hawai'i, dolphin swim tourism is viewed by some as a positive tool for education, conservation, and economic growth, while others feel that human activities are having negative impacts on the spinner dolphin (*Stenella longirostris*; see Figure 8.1) population, affecting their ability to feed, reproduce, and socialize (Würsig, 1996; Östman-Lind et al., 2004; Danil et al., 2005; Bejder et al., 2006; Delfour, 2007; Courbis and Timmel, 2009). Dolphin tourism in Hawai'i is a profitable business and many people are dependent on the industry. Dolphin swim tourism conflict between local communities, operators, and government has occurred as a result of fishing ground disputes, ethical and legal debates, and economic interests. Different types of dolphin swimmers who diverge in their attitudes, reasoning, and frequency of

Figure 8.1 Hawaiian spinner dolphins (*Stenella longirostris*). Julian Tyne, NOAA #GA15409

dolphin swims also play into this clash. Each group holds distinct values and opinions regarding the dolphins and related management conservation plans. Many groups' identities are tied to their values and beliefs about the dolphins. The conflict, while focused on dolphin swims, does not occur between the dolphins and community but rather is flanked by different interests humans have in wild spinner dolphins. As for other human–animal conflicts, the source of tension in Hawai'i stems from deeper issues between community groups.

This chapter will present a unique examination of the dolphin–human interface using popular culture within a tourism context. Common characteristics of dolphins as free, intelligent, and harmonious animals are rooted in idealized thinking about cetaceans (Anderson and Henderson, 2005; Fraser et al., 2006). These beliefs continue to be present in popular representations, emphasizing strong emotional bonds between dolphins and humans. Cultural depictions of dolphins in media, arts, and other forms will be presented using six common personas often assigned to dolphins. Links between mediated dolphin narratives and the growth of dolphin swim tourism will be made in relation to community conflict and changes that come with tourism expansion. The levels of conflict model (Madden and McQuinn, 2014) will frame this case study, exploring how the act of swimming with dolphins has changed community composition and interests and serves as a symbolic manifestation of other societal conflicts.

8.3 Mediated narratives of dolphins

The wild dolphin swim experience revolves around expectations of dolphin performance gathered through familiar images. Exploring popular sources such as literature, film, television, visual arts, and other cultural forms is fundamental to human perceptions of care and concern toward animals, playing an important ideological and intellectual role in Western thought (Fudge, 2002). Common characterizations of the dolphin as wild, intelligent, altruistic, happy, and alien animals are deep-rooted in thinking about dolphins, but how precise are these portrayals? Cloke and Perkins (2005) contend that these representations may not be accurate, as humans tend to divorce social associations from actual animal behaviors. Depictions of animals in media are blurred between portrayals of the real and simulated, influencing moral boundaries in humans (Fudge, 2002). Anthropomorphic social representations of dolphins are created to fit human needs.

The promotion of dolphins as creatures that are wild and different from humans is explored in science fiction novels and movies (Bryld and Lykke, 2000). Contrariwise, dolphins are also interpreted as being similar to humans, possessing characteristics such as intelligence, advanced language skills, altruistic behavior, and play. Desmond (1999) identifies this conflicting tension as defining dolphins as outliers or boundary crossers straddling the lines between humans and fish. Six common constructions of the dolphin figure portrayed in media and literature have been identified and contrasted to determine how these depictions fit into human values imposed onto dolphins (see Table 8.1).

Table 8.1 Common constructions of the dolphin figure found in media and literature

Dolphin figure	Attributes	Description	Examples Film and television	Literature
Dolphins as aliens/science fiction dolphins	Extraterrestrial, other worldly, futuristic	Body modification, animal as machine, science fiction	*Johnny Mnemonic* (1995) *SeaQuest DSV* (1993–1996) *The Day of the Dolphin* (1973) *The Hitchhikers Guide to the Galaxy* (2005)	*A Deeper Sea*, Alexander Jablokov *Cachalot*, Alan Dean Foster *Deep Wizardry*, Diane Duane *Hyperion*, Dan Simmons *Johnny Mnemonic*, William Gibson *Known Space*, Larry Niven *So Long and Thanks For All The Fish*, Douglas Adams *The Scar*, China Mieville *The Uplift Series*, David Brin
Dolphins as intelligent	Telepathic, superior powers, hero	Key to the universe, viewer of earth and other worlds, communicates with humans	*Dolphins* (2000) *Eye of the Dolphin* (2006) *SeaQuest DSV* (1993–1996) *The Day of the Dolphins* (1973) *The Hitchhikers Guide to the Galaxy* (2005)	*A Ring of Endless Light*, Madeline L'Engle *Cachalot*, Alan Dean Foster *Dolphin Island*, Arthur C. Clarke *So Long and Thanks For All The Fish*, Douglas Adams *Startide Rising*, David Brin *The Dolphins of Pern*, Anne McCaffrey
Dolphins as performers	Entertainer, jester, happy, trickster	Enjoys the spotlight, showing off, always cheerful and ready to play	*Dolphin Tale 1* (2011) *Dolphin Tale 2* (2014) *Flipper* (1964–1967) *Free Willy* (1993)	

Table 8.1 (*continued*)

Dolphin figure	Attributes	Description	Examples	
			Film and television	**Literature**
Dolphins as caretakers	Maternal, loyal friend, hero, unconditional love, altruistic	Associated with healing powers, mystical/angel-like, protector of humans, saving people from drowning, guardian dolphins, fishing village dolphins	*A Ring of Endless Light* (2003) *Dolphin Boy* (2011) *Dolphin Cove* (1989) *Dolphin Tale 1* (2011) *Dolphin Tale 2* (2014) *Eye of the Dolphin* (2006) *Flipper* (1963, 1996) *Flipper* (1964–1967) *Free Willy* (1993) *The Day of the Dolphin* (1973)	*A Ring of Endless Light*, Madeline L'Engle *Deep Wizardry*, Diane Duane
Dolphins as boundary crossers	Neutral gender, human-like, once lived on land and now in the ocean, captive and wild	Mimicry, shapeshifter, straddles the line between human attributes and wild animal, talking dolphins	*Dolphin Tale 1* (2011) *Dolphin Tale 2* (2014) *SeaQuest DSV* (1993–1996) *The Day of the Dolphin* (1973)	*Deep Wizardry*, Diane Duane *Ishmael in Love*, Robert Silverberg
Dolphins as wild	Pastoral, exotic, free, manifestation of miraculous, innocence	Idealized or romantic view of nature, sexualized/sexual beings	*Beneath the Blue* (2010) *Black Fish* (2013) *Dolphins* (2000) Nature documentaries *The Cove* (2009)	

8.3.1 Dolphins as aliens

The concept of dolphins as extraterrestrials has been considered by several authors, including Lilly (1961), Nielsen (1994), Ocean (1997, 1989), and White (2007). This concept is based on both a literal and metaphysical definition of aliens and the parallels between the unknown ocean and unexplored space. The concept of dolphins as extraterrestrials highlights their immense differences from humans (as animals), existing in a foreign aquatic environment (White, 2007). The dolphin as an "exotic other" has literally led to the belief by some that dolphins are aliens from another planet and have come here to share insights into the human mind (Ocean, 1989; Bryld and Lykke, 2000; Servais, 2005). Some of the authors who subscribe to this belief were inspired by John Lilly, a 1950s neuroscientist who helped promote this idea when he played recordings of dolphins "speaking English" at a conference on extraterrestrial intelligence (Bryld and Lykke, 2000). Following his presentation in 1961, Lilly expanded his dolphin communication research and worked on computer-aided interspecies language development with two captive dolphins, Joe and Rosie (Cochrane and Callen, 1992).

Following the expansion of dolphin–human communication research in the 1960s, several groups in North America emerged in the early 1980s citing the ability to telepathically communicate with dolphins. One of the first groups was founded in Hawai'i and claimed to have explored dolphin spirituality by swimming with dolphins and establishing telepathic exchange with them (Ocean, 1997, 1989). The dolphins supposedly revealed to this group that they were aliens from another planet who were sent to earth to help humans live a more peaceful and natural life (Ocean, 1989). The same philosophy has been adopted by several other groups internationally, who have published books on a similar theme, including *Mind in the Water* (McIntyre, 1974).

The perception of dolphins as mystical guides with superior intellects is exemplary of the romantic notion of dolphins as spiritually and intellectually enlightened beings (Fraser et al., 2006). The literal concept of dolphins as aliens has also become prevalent in certain circles, leading to groups of healers, authors, and therapists pushing for interspecies cohabitation and relationships, touting the abilities of dolphins to directly communicate with and heal humans (Servais, 2005). One of the central locations for these groups to gather and work with dolphins is on Hawai'i Island. Many community members offer retreats, water birthing, and healing opportunities, working with resident spinner dolphins. Their unmistakable presence in Hawai'i has grown in communities close to the bays where dolphins frequent, with advertisements for retreats and gatherings.

Lilly's influence on interspecies communication has continued both in groups who practice it and in the realm of science fiction storytelling. The interplay of reality and fiction can sometimes become blurred as dolphins occupy the cultural imagination. Lilly's work and association with dolphins has trickled into popular culture references, as demonstrated by the unusual number of dolphins used in science fiction novels and movies (e.g., *Startide Rising* (1996), *Johnny Mnemonic* (1995), and *Day of the Dolphins* (1973)). From psychic healing and spiritual guidance to dolphin robots that speak English, dolphins have been a popular source of miracles and futuristic plot devices since the 1963 film *Flipper*. In addition, dolphins are often highlighted for their extreme intelligence (Sandoz-Merrill, 1999; White, 2007; Warren, 2009).

8.3.2 Dolphins as intelligent beings

Large brains and storied intelligence have led to the common association of dolphins as being one of the most intelligent nonhuman animals. Some dolphins have shown that they are aware of themselves and their mental processes and have demonstrated the capability to display abstract concerns and process creative impulses (Griffin, 1986; Bekoff, 2002; Herman, 2002; Simmonds, 2008). These attributes contribute to the idea of dolphins as being completely sentient or emotionally conscious creatures, an idea which has been constructed through many social relations between dolphins and humans (Peace, 2005). Authors such as Ocean (1989) and Nielsen (1994) credit dolphins not just for being intelligent but also for possessing a conservation-minded understanding of the natural world. Because of this, dolphins are viewed as key to the universe's struggles, with abilities to solve problems and heal humans. This expansion of the dolphin's intellect is perpetuated in fictional literature such as *Dolphin Island* (1963) and *Startide Rising* (1983) and in films like *Eye of the Dolphin* (2006) and *The Hitchhikers Guide to the Galaxy* (2005). The portrayal of dolphin intelligence in these works has further led to the general public's desire to see these supposed capabilities firsthand. There are few animals who demonstrate similar characteristics in ways that are visible to humans, and the association of dolphins with a unique intelligence has intensified people's longing to connect with them.

8.3.3 Dolphins as performers

Their intelligence has also made dolphins desirable for use as performing animals, as they are trainable and can execute a varied repertoire of behaviors. One of the most common initial interactions people have with marine mammals is through aquariums that have performing dolphins. Dolphinariums throughout the world have developed the persona of dolphins as entertainers. Images of captive dolphins as happy and ready to play are promoted by such facilities, contributing to the commonly perceived dolphin identity of joker, trickster, and comic. The image of dolphins as vehicles for entertainment has perpetuated the dolphin swim industry, both in captive facilities and in the wild. The perception of dolphins as natural entertainers strongly influences participants' expectations during wild encounters. In locations where both captive and wild dolphin swims are available, there can be confusion among human participants as to what dolphins will do during the meeting. Spectacular images via an array of virtual outlets shape how humans feel about dolphins and their expected abilities (Newsome et al., 2005). Even before someone gets on a boat, the prospect of establishing social relationships with dolphins is anticipated (Peace, 2005). The perception of the dolphin as a performer is reinforced by translations of dolphin behavior into human terms, as articulated in popular films such as *Flipper* (1963) and nature "documentaries" and perpetuated by dolphinarium trainers, wild dolphin swim guides, and captains of boats promoting wild dolphin swims.

8.3.4 Dolphins as caretakers

Intelligence and performance are not the only themes that permeate the popular perception of dolphins. Another commonly assigned attribute is that of caretaking or saving, which relates to dolphins' storied protection of humanity. People who have encountered dolphins in the open sea regularly speak about falling in love with or feeling love from dolphins; others report telepathy, trances, or mystic revelations. The emotions that humans experience in a dolphin encounter are very powerful. Dolphin spiritualism gives voice to the idea that dolphins have the ability to improve human physical and emotional well-being, creating a following of people who participate in dolphin healing or therapy. Many report the "dolphin effect" after a dolphin swim, testifying to feelings of change or healing, including the alleviation of chronic depression, the removal of pain, and recovery from illness (Cochrane and Callen, 1992; Selke, 1997; Taylor, 2003). Burnett (2010) explains that supporters of the "dolphin effect" believe that dolphins have the ability to "reach inside" the human body for diagnostic or palliative purposes.

The idea of dolphins aiding or healing humans is not new. Dolphins' legendary helpfulness has been reported in histories of dolphin–human encounters. Incidents of dolphins saving people from drowning have been described in stories by Aesop, Herodotus, and Pliny the Elder and mentioned by Shakespeare (Bulbeck, 2005). Across the Pacific, there are fables of enduring friendships between dolphins and sailors, of dolphins helping fishermen, and of dolphins rescuing shipwrecked humans (Cressey, 2009). These stories of rescue and guidance have helped perpetuate the dolphin as a popular figure of love in film, television, and literature. The theme of mutual caretaking is also common, where someone rescues a dolphin from a dangerous situation and in return the animal ends up saving the human (e.g., *Dolphin Tale* (2011), *Eye of the Dolphin* (2006), and *Free Willy* (1993)). Another popular rescue plot revolves around dolphins' healing powers to save a sick or misunderstood person (e.g., *A Ring of Endless Light* (1980) and *Deep Wizardry* (1985)).

8.3.5 Dolphins as boundary crossers

The idea of dolphins as boundary crossers is based on the fact that they were land inhabitants who transformed into ocean dwellers, and the observation that they possess both animal and human-like characteristics. Bulbeck (2005) argues that dolphins exist in the borders of human classification because they represent contradictions by being "brainy sea mammals" and "fish with brains." The mystery of dolphin lives has played into folklore across the globe, associating dolphins with magical powers and the ability to merge with people on land. For example, in South America some people view the river dolphin (or boto) as a mysterious being that can shape shift and capture souls (Montgomery, 2003).

Perceptions of dolphins as boundary crossers are based on the dolphin's ability to inhabit an unknown aquatic universe while still being somewhat accessible to humans. Other dualities include the dolphin's need to breathe air despite spending an extended time in water, and the ability to explore their own world with both sound and sight (Bryld and Lykke, 2000).

Many films and television shows portray dolphins as intelligent beings that save the day or significantly contribute to the discovery of a solution to an important problem. At the same time, these protagonists are kept in captivity and treated as sidekicks or pets of the human characters. Examples of this can be seen in the television show *SeaQuest DSV* (1993–1996) or the film *The Day of the Dolphin* (1973).

The dolphin figure is also known as a vehicle that can span human boundaries through various modes of interspecies communication, such as telepathy or other forms of non-verbal communication, or via the Dreamtime. In many cultures across the world, dolphins have been viewed as ancestors or family protectors of people, straddling the line between animal and spirit. Several shamanic and tribal cultures perceive humans and animals as different manifestations of the same spirit, easily interchanging with one another. In Native Hawaiian traditions some consider the dolphin to be *amakua* (family guardians), while the people of Northern Australia believe the bottlenose dolphin to be their totem (Nollman, 1985). Spiritual and mythical dolphins communicating with humans are portrayed in the stories of many indigenous cultures; for example, see the Chumash legend on the origin of dolphins (Santa Barbara Museum of Natural History, 1991), and the legend of Katama (Hall, 2004).

8.3.6 Dolphins as wild animals

Humans have also understood dolphins as wild animals free from inhibition, sexually unrestricted, and acting on primal needs. The construction of dolphins as autonomous "others" with free liberties is romanticized in popular culture. Besio et al. (2008) describe dolphins as bearers of alternative values, representing "indescribable otherness" that captures and fascinates humans. Sexuality has been the most common focus of dolphin "otherness," based on their promiscuity with opposite- and same-sex partners. This ambivalence concerning mates has been captured by Brazilian folk tales of botos tricking village girls into having sex, thus impregnating them (Montgomery, 2003; Slater, 2004).

Besides being linked to sexuality, dolphins are seen through a romanticized lens to represent nature, offering a free and utopian society far greater than that provided by humanity (Davis, 1997). Portrayal of this is common in nature shows and documentaries. The fact that dolphins live in the ocean, one of the last remote and unexplored areas of earth, further reinforces the "dolphin mystery" and human awe of this species.

8.3.7 Mediated dolphin narratives drive dolphin swim popularity

Newsome et al. (2005) and Evernden (1999) argue that human values influence how people view and treat animals, providing a set of beliefs about the importance of wildlife. These assigned characteristics ultimately create a construction of dolphins as a resource for entertainment, as a peer of equal intelligence, or as a miraculous being with spiritual and superior qualities to humans (Evernden, 2003). The subtle influence of "dolphin culture" in the media and literature has perpetuated global dolphin swim tourism, shaping human expectations of encounters with wild dolphins. Those seeking dolphin swim experiences are looking for the connections they have seen in media, whether for entertainment, healing, or living out a life-long dream. The Hawaiian Islands provide an interesting context for wild dolphin swims, as they attract participants looking to achieve all three of these goals. Not only has this perpetuated the exponential growth of wild dolphin swim operators and tours, but it has also created a divide among the community between those who are in support of swimming with the dolphins and those who are not. Hawai'i is not the only location where this global activity is occurring; however, it is one of the first places where wild dolphin swims began, and it has a 40-year history of the practice.

8.4 Hawaiian spinner dolphin tourism: a case study

Hawai'i is home to the Hawaiian spinner dolphin (*Stenella longirostris*), a small marine mammal known for its aerial spinning and jumping behavior. Unlike other dolphins, spinners forage at night and rest during the day, making their presence in sandy bays predictable and easy to spot (Norris et al., 1994; Tyne et al., 2014). In the 1970s some residents began swimming with the dolphins, taking note of the ecstatic feeling they would get after participating in swims (Wiener, 2014). Slowly this transformed into guided tours with specialized groups looking to connect with the dolphins on a spiritual level. As the number of spiritual

Figure 8.2 A typical setting in Kona where swimmers attempt to interact with Hawaiian spinner dolphins. Julian Tyne, NOAA #GA15409

swimmers continued to grow in size, the activity began to expand to the commercial sector, with a few tour boats devoting their business to swimming with dolphins (Figure 8.2). As their success grew, so did the number of boats and the popularity of this activity.

As of 2014, there are two locations in the Hawaiian Islands where visitors can participate in wild dolphin swim tours. On Hawai'i Island (Big Island) there are also bays where people will swim unguided with the dolphins from the shore. As the activity has expanded, communities living near the bays where dolphins frequent have become increasingly worried about the growth in traffic, the use of the area, and the impact on the dolphins. This concern has been echoed by government workers charged with managing local marine resources. Additionally, some in the Native Hawaiian community have spoken out against dolphin swims, while others have embraced it. The polarizing concerns associated with the local dolphin swim industry have not only created controversy about the activity but have generated deep divisions within the communities as well.

8.4.1 Dolphin swims and community conflict

The community conflict surrounding spinner dolphins includes four primary issues: (1) governance of common resources; (2) tourism contributions to economic sustainability; (3) cultural identity; and (4) social relationships (Silva, 2013). The role of spinner dolphins in the community and what dolphins mean to each individual involved in the conflict is central to the ethical division among involved parties (see Comment). A dualism exists between stakeholders, based on how they conceptualize human relationships and behavior toward dolphins. Some view the dolphins as superior to humans, affording agency to dolphins to make their own decisions about swimming with humans. Others argue that the dolphin swims should not occur out of respect for the community's cultural values (discussed in Section 8.4.2.3) and for the safety of the species (Wiener, 2014). Conflict also exists between community members and those specifically profiting from dolphin swims, as the latter argue that the swims are floating the local economy. At the heart of the spinner dolphin swim discord is the pending increase in government regulation limiting or controlling access to interactions with dolphins.

Stakeholder engagement that addresses deeper drivers of the disagreement is typically not considered in human–wildlife conflict, and therefore underlying differences go unaddressed (Madden and McQuinn, 2014). This happens because hidden social struggles are not analyzed and transformed through stakeholder engagement processes (Madden and McQuinn, 2014). Four stakeholder groups lie at the root of conflict surrounding Hawaiian spinner dolphin swim tourism: commercial dolphin swim operators; residents who are against dolphin swims, including most fishermen/women and most of the Native Hawaiian community; residents who swim with the dolphins (noncommercially); and the marine management agencies and scientists. Ironically, all stakeholders believe they are doing what is best for the animals and are genuinely concerned with their welfare. This is a commonality that should be built upon when examining the deeper connections between the disagreeing parties.

> **Comment**
>
> To see how this plays out on an international scale, go to E. C. M. Parsons's depictions in Chapter 5 of how competing cultural interpretations of whales has exacerbated conflict at the IWC.
>
> —*The Editors*

8.4.2 The levels of conflict model

The levels of conflict model provides a means to assess the complexity, scope, and depth of the dolphin swim tensions in Hawai'i. The levels of conflict model acts as a guide to flesh out the contentious history surrounding this dispute. Each stakeholder has their own meanings and set of emotions tied to the dolphin swim issue, and these go deeper than whether it is okay to swim with wild dolphins. Madden and McQuinn (2014) have designed a process that closely examines relationships surrounding wildlife conflict and provides a means for reconciliation by humanizing all participants. This process works to break down the "us" versus "them" mentality that often accompanies conflict (Madden and McQuinn, 2014). A long-standing history of inequity has existed in the Hawaiian Islands and provides the context for deeper issues underlying the dolphin swim debate. Three levels of conflict will be reviewed based on the various stakeholder groups and their interests. These groups will be further divided into those who support a rule change banning or limiting swimming with dolphins, and those who do not.

8.4.2.1 Dispute conflict

The first level, dispute conflict, is typically the surface-level conflict. This is where people debate whether the public should be allowed to swim with the wild Hawaiian spinner dolphin population. While the question of whether to regulate dolphin swims may seem like a cut-and-dry issue, it is far more complex (see Table 8.2). The two groups in favor of wild dolphin swims are the operators and noncommercial swimmers. The operators are people who financially benefit from the swims by owning companies, working on boats leading dolphin tours, or organizing group visits to the dolphins (Wiener, 2014). Their greatest concern is the loss of income from a possible dolphin swim restriction suggested by federal marine management agencies. Although commercially driven, this group is concerned about the dolphins' welfare and has expressed worry over the increasing number of boats and businesses (Wiener, 2014).

Those who swim with the dolphins for noncommercial purposes are also nervous about the potential restrictions to dolphin swims. These residents typically swim with the dolphins several times a week and are proficient-enough swimmers to dive alongside the dolphins when approached and engage them on closer levels than visitors who have never swam with a dolphin before (Wiener, 2014). This is a tight-knit group who rely on each other for reports as to where the dolphins are located. The majority of them feel that it is their right to swim with

Table 8.2 Dispute conflict for stakeholder groups associated with wild dolphin swims

Against dolphin swims		For dolphin swims	
Federal marine management agency and scientists	Residents against dolphin swims; Native Hawaiian community; fishermen/women	Dolphin swim operators	Noncommercial dolphin swim residents
• Current concern for the growth in dolphin swim tourism, the commercial operations, and the population effects this may have on the Hawaiian spinner dolphin.	• Impact of dolphin swimmers on communities near dolphin bays (increased traffic, beach use, etc.). • Current concerns over the dolphin's well-being and population.	• Against proposed dolphin swim restrictions by federal marine management agencies due to potential loss of livelihood/income. • Uncontrolled increase of industry (too many boats, competition, etc.).	• Against proposed dolphin swim restrictions by federal marine management agencies due to potential loss of recreation and enjoyment of the dolphins. • Impact of visitors on dolphin swim beaches via crowding the areas that they swim in.

the dolphins and that they are not impacting the population. The swimmers have also vocalized concern for the dolphins and feel that the growing boat traffic and commercial groups are where the problems with human–dolphin interactions lie (Wiener, 2014). Many complaints about the increased traffic in the water caused by boats and tourists have been made as the popularity of this activity continues to grow.

The stakeholder groups not in favor of dolphin swims are typically marine managers and scientists, as well as community residents who do not swim with the dolphins. This group would like to see federal regulations change the way people currently interact with dolphins. Community residents who live near the bays where dolphins come to rest sometimes participate in subsistence or commercial fishing and may be part of the Native Hawaiian community. They do not engage with the dolphins and are upset with the amplified traffic and beach use due to the dolphin swims. This group, as well as the federal marine management agencies and most scientists, are concerned about tourism growth affecting the spinner population.

On the surface, the main question of whether dolphin swims should be allowed seems to be the dividing factor between the different sides of this conflict. However, this is not really the case. Even those who share the same beliefs about dolphin swims are divided on how this should be handled. For example, some community members who think of dolphins as mystical, wild friends may choose not to engage in swimming with dolphins to protect them, while others with similar conceptualizations would attribute the behavior of swimming with dolphins as an appropriate manifestation of dolphin friendship. Marine managers cannot make any decisions to restrict dolphin swims without considering all perspectives and factors at play within the different communities, including the finances for enforcement, the impact of restrictions on other activities near the dolphins such as fishing, and the way people engage with the dolphins themselves (shore vs. boat).

8.4.2.2 Underlying conflict

The second level of conflict sheds light on the next layer of the dolphin swim community struggle, providing context for why people feel the way they do. Underlying conflict results from past decisions or unresolved interactions that intensify or aggravate the present situation (Madden and McQuinn, 2014). Underlying conflict adds significance to current disputes that are not necessarily obvious from the bare "facts" alone (Madden and McQuinn, 2014). Similar to the dispute or surface conflict, each stakeholder may have a different history and therefore varied reasons for their stance. Even within a particular group, individual experiences will change, altering the intensity of their feelings toward the same issue (see Table 8.3).

Since the dolphin swim conflict has been ongoing since the 1980s, the government has stepped in to try and manage the issue on numerous occasions. Over the past ten years, more aggressive steps have been taken to change rules; however, the slow process and looming threat of regulation has left operators disillusioned and hostile toward marine managers (Table 8.4; Wiener, 2014). Community members who are in support of restricting dolphin swims have also experienced this drawn-out process, which has added to the already negative feelings toward the government and scientists who have promised rule changes that have

Table 8.3 Underlying conflict for stakeholder groups associated with wild dolphin swims

Against dolphin swims		For dolphin swims	
Federal marine management agency and scientists	Residents against dolphin swims; Native Hawaiian community; fishermen/women	Dolphin swim operators	Noncommercial dolphin swim residents
• Discouraged over failed attempts to work with the community in the past to implement responsible business practices. • Frustrated by internal government time restraints to implement proposed actions.	• Frustrated by inaction or length of time taken by federal and state government over dolphin conservation/management.	• Years of government mismanagement of dolphin swim industry has led to irritated operators.	• Years of government mismanagement of dolphin swim industry has led to increased commercial vessels crowding swim areas, resulting in irritated swimmers and concerns for the dolphins from growing commercial industry.

Table 8.4 Timeline of the dolphin swim industry in Hawai'i

Time Frame	Activity
Late 1970s	No formal or daily dolphins swims were established, but small groups camping near bays would swim out to dolphins on occasion when they were close to the shore.
1989	Dolphin swims started becoming more common in Kealakekua Bay on Hawai'i Island. One person began taking out small groups of friends and then others that were interested, slowly developing the first dolphin swim business.
1996	Upon seeing the success on Hawai'i Island, the first two companies began in Wai'anae, O'ahu. The companies used small vessels including a sailboat and zodiac.
1997/1998	Two more companies begin operations out of private harbor facility on O'ahu.
1998	Gentleman's agreement to limit commercial boat operations to seven boats on O'ahu.
2000 onward	Mass expansion of tour operators on both O'ahu and Hawai'i Island.
2005	U.S. federal management agency, NOAA, releases advancement notice of proposed rule pertaining to dolphin swim tourism.
2006	Statewide public scoping meetings and public comment period on the proposed rule.
2010	The Coral Reef Alliance establishes community voluntary standards for dolphin operators.
2011	Dolphin SMART responsible dolphin tour recognition program introduced in Hawai'i.
2014	NOAA initiates draft Environmental Impacts Statement of dolphin swim tourism and completes a Cultural Impacts Assessment.
2015	NOAA initiates draft rule writing with plans to release for public comment sometime in 2015.

not yet come to fruition (Wiener, 2014). Many community meetings have been held in both O'ahu and Hawai'i Island (Big Island) allowing for public comment. The problem is that rarely does the community feel that the government has addressed or considered their comments when developing potential rules.

Public hostility does not just stem from the dolphin swim issue but also from other community problems, including manta ray dive tourism, fishing regulations, and management of coastal waters. Past conflict has been so intense that several individuals and nonprofit groups working on the dolphin swim issue have abandoned the cause. Changeover in government workers running the meetings, together with a history of similar topics being addressed with no lasting changes, has perpetuated anger within the community (Wiener, 2014). In the meantime, "no official rules" have been issued, and increased vessel crowding in swimming and fishing areas has become a source of contention, leading to several physical fights between operators and community members. The community feels disenfranchised by the slow timescale and potential of top-down decisions being made on resources that they depend on for sustenance, recreation, and cultural purposes.

Marine managers and scientists have been equally frustrated as, despite their hard work, the community believes they are not making an adequate effort, and they are routinely painted as "bad guys" who do not care about the community (Wiener, 2014). Managers have been pushing for new regulations for many years but have been delayed in their efforts. The managers are also discouraged by the long time frames necessary to produce rule changes. Managers have been able to make inroads with some operators and industry members but for the most part feel that the community will not work with them and are not willing to take small steps to self-regulate their behavior (Wiener, 2014). This does not apply to all operators; there are so many companies that attitudes and behaviors differ between companies and islands (Wiener, 2014). This makes it extremely difficult for managers to implement a statewide policy. One self-regulation attempt has been made using the dolphin SMART program, an operator recognition program offered by the federal marine management agency NOAA and its partners. Unfortunately, this program recognition has caused tensions among operators, as it only endorses companies that do not swim with dolphins. Some operators have viewed this program as an attempt to criminalize legitimate businesses and argue that dolphin-watching companies have the same potential to harm dolphins as dolphin swim companies (Wiener, 2014).

8.4.2.3 Identity-based or deep-rooted conflict

The final conflict level, identity-based or deep-rooted conflict, involves the values, beliefs, or social-psychological needs that are central to the identity of at least one of the parties involved and are perceived to be threatened by another identity group (Madden and McQuinn, 2014). These types of conflict go unaddressed and are often ambiguous and difficult to respond to (Madden and McQuinn, 2014). Given the polarized identities related to a long history of resource-use conflict in the Hawaiian Islands, dolphin swim issues are not the only problems stemming from deep-rooted conflict. Community conflict is not limited to cross-stakeholder group conflict, such as between operators and marine managers, but also exists within stake-holder groups such as individual dolphin swim operators or companies (see Table 8.5).

Differences can be seen between operators based on which island they work from, how long they have been operating, and the type of business that they attract (Wiener, 2014). For instance, companies on the island of O'ahu own larger boats that hold more people than those owned by operators on Hawai'i Island. However, O'ahu operators do not allow their participants to dive with the dolphins and make them wear life jackets (Wiener, 2014). On Hawai'i Island, most operators have smaller boats that provide a more intimate experience (Wiener, 2014). Since the groups are smaller, there are more boats, and participants are free to swim with the dolphins without flotation devices (Wiener, 2014). On both islands, operators who have been in business the longest have close relationships and work together to find the dolphins; long-standing operators are skeptical of some of the new companies who do not follow the established etiquette in the same way (Wiener, 2014). Some of the older operators

Table 8.5 Identity-based conflict for stakeholder groups associated with wild dolphin swims

Against dolphin swims		For dolphin swims	
Federal marine management agency and scientists	Residents against dolphin swims; Native Hawaiian community; fishermen/women	Dolphin swim operators	Noncommercial dolphin swim residents
• Those who are in favor of regulations are frustrated by the prejudice from dolphin swimmers and operators automatically assuming that the federal government does not care about the community. • Operators are frustrated with managers assuming that they do not care about the dolphins. These long-time opposing feelings have led to deep entrenched individual and groups positions, fueling the conflict. • Government officials are repeatedly viewed as "the bad guy," even as new hires come in who have no history working in the area/on this issue.	• Loss of way of life from the inability to use dolphin bays for subsistence fishing due to dolphin swims, increased traffic. • Lack of respect for culture that stems from a long history of Native Hawaiian repression from the federal government. • A strong history of government mistrust leading to anti-legislation. • Negative feelings toward non-Hawaiian people moving to the area for the dolphin swims, stemming from a long history of perceived land rights abuses.	• The operators feel unfairly targeted by the federal management agencies and are concerned that they hold an anti-business sentiment. Concerns about the government using their power to shut down their livelihood. • The operators feel that they care for the dolphins and are not harming them. Many operators feel personal connections with individual dolphins and have long-term relationships with them. The operators are insulted by accusations that as a group they are responsible for hurting the dolphins.	• Dolphin swim residents who believe in dolphin healing and telepathic abilities feel that their spiritual practice and way of life is being threatened and that no one has a right to take this practice away from them. Other community members who do not believe in these practices label and ostracize this group for their beliefs of spiritual connections to the dolphins.

hold prejudices toward the newcomers because the latter run their boats differently from the former and create competition. The long-standing operators are mostly in favor of some changes to the regulations but would rather see a cap on the number of companies allowed than a ban on dolphin swims (Wiener, 2014). The operators feel unfairly blamed for the possible decline in the dolphin population and do not believe there is much evidence to support marine managers' claims that the dolphin swim companies are the sole cause of the decline.

Noncommercial dolphin swim residents also have different feelings about this issue, ranging from those who swim as part of a spiritual experience to those who participate recreationally. A large group of swimmers who have religious and spiritual ties to the dolphins feel that potential law changes would threaten their identity as people who help the dolphins.

This group believes that by swimming with the dolphins they are able to enrich their lives and that the dolphins make their own choices to interact with humans (Wiener, 2014). Swimmers who do not swim for spiritual purposes often get lumped with those that do. Both groups feel that the marine managers are unfairly targeting dolphin swimmers because of their beliefs and practices and that residents should be exempt from regulations related to commercial dolphin swims (Wiener, 2014). It is important to keep in mind that these stakeholder groups do not have clearly drawn borders, and relationships between them can be blurred. For example, noncommercial dolphin swimmers will sometimes work with commercial operators or befriend those who swim with dolphins for spiritual practice.

Residents who are not in support of dolphin swims feel that their way of life is being threatened as a byproduct of wild dolphin swims. This belief is especially strong among the Native Hawaiian community and fishermen/women, who have already dealt with a long history of rights and land abuses. The fishermen/women have seen a rapid decline in their fish stock and changes in the rules for recreational and subsistence fishing. Those who fish for *akule* (big-eyed scad; *Selar crumenopthalamus*) are particularly vocal in attributing the loss of fish to dolphin swim tourism. *Akule* was a staple fish in the Hawaiian diet and a favorite for drying (Info Grafik, 2014). *Akule* are found in shallow waters similar to those found in the resting habitat of dolphins, and some in the fishing community believe that sound from the dolphin swim boat motors has partially driven the fish away (Wiener, 2014). Several people within the Native Hawaiian community have also said that the dolphin swim activities do not respect Native Hawaiian culture and that the dolphins are not there for leisure activities (Wiener, 2014).

The combined history of mistrust and lack of control over resources has perpetuated fear of and resentment toward government involvement in handling the dolphin swim issue. Even if government-imposed regulations would help decrease the number of boats and people swimming with the dolphins, many in the community would not trust top-down regulations because these rules might impact other activities such as fishing (Wiener, 2014). There is also a history of government mistrust among the Native Hawaiian community that goes back to the U.S. government takeover of Hawai'i. The Native Hawaiian community would prefer to self-manage their own bays so that they can control the rules impacting their activities and practices (Wiener, 2014). A deep history of inequitable land rights has also caused hostility toward dolphin swim participants, who tend to be Caucasian females from the U.S. mainland (Wiener, 2014).

A deep-rooted conflict within the community exists between those who are from Hawai'i and others who have moved to the islands, including those from the U.S. mainland. This remains consistent in dolphin swim conflicts, producing deeper resentment for the groups of dolphin swimmers who typically are not from Hawai'i but come and use Hawaiian resources. The dolphin swim conflict is similar to many other conservation battles that typically serve as platforms for deeper concerns and unmet needs such as empowerment, identity, and spiritual security (Madden and McQuinn, 2014). A long-standing history of foreigners taking over Native Hawaiian land and resources has fueled these feelings toward dolphin swimmers. This history has also perpetuated an attitude toward some of the government workers trying to establish regulations in the communities. New employees or scientists who come to work on dolphin issues are met with resistance, even when they have no previous involvement or history in the community (Wiener, 2014).

The act of swimming with dolphins itself goes against some cultural rules such as not entering a different community's bay unless given permission (Wiener, 2014). For these reasons, dolphin swims are perceived as threatening the community's identity and way of life. This mistrust generated by the Native Hawaiian community is targeted toward the noncommercial resident swimmers as well as the federal management agencies and scientists.

8.5 Conclusions

The controversy surrounding swimming with dolphins is just one example of the challenges that face human–wildlife relationships. Dolphin swim tourism continues to show an upward trend, particularly with the continued coverage of dolphins in pop culture, as described earlier in this chapter. Since human interest in dolphin tourism shows no signs of letting up, solutions to dolphin swim tourism community conflict and animal well-being must be considered.

Media has the power to influence how people view and interact with wildlife. The recent orca-captivity documentary *Black Fish* (2013) has illustrated the influence film can have in altering public perception. As demonstrated earlier, the portrayal of dolphins in the public forum has not changed much in the last 50 years. The fantasy of outlandish, intelligent, and caring dolphins still floats in the public's mind when engaging with them. The act of imposing human-centered attitudes on the natural world affects decision-making toward nonhuman species (Fawcett, 1989; Wilson, 1993). For example, some assume that dolphins are always happy because of their large permanent "smile," but this is due to the structure of their jaw line. This perceived happiness can affect decision-making, as people assume that when dolphins are smiling they are enjoying human interactions. Expectations for interactions with dolphins need to be managed and realistically represented in order to provide a safe environment for people to view dolphins and treat them with respect, while still allowing for communities to continue with existing tourism.

Analyzing the three levels of conflict as they pertain to Hawaiian spinner dolphin tourism provides a background to the complexity and severity of the dolphin swim issues at hand and the related community context. By identifying the levels of conflict among the four stakeholder groups discussed in this chapter, it is clear that there are a number of obstacles to a simple resolution, and given the intensity and complexity of the conflict, current methods and approaches for decision-making have not been effective. A transformative process may help to create the conditions for shared problem-solving and mutual respect that may alleviate some of the long-standing identity conflicts and provide an environment where communities can connect via their mutual admiration for the Hawaiian spinner dolphin (Madden and McQuinn, 2014).

Dolphin swim conservation management has to focus on all factors including physical and spatial measures, economics, history, legal actions with enforcement, and most importantly, the psychological, social, and cultural values and needs of all stakeholders (Madden and McQuinn, 2014). What makes each stakeholder group so passionate is that all individuals involved have a different experience with the dolphins, an experience that is deeply personal. As John Livingston (1994) said, "Nature is entirely subjective and value-laden"; it is for this reason that all parties must be felt that they are heard and valued through a more transformative process of dialog and decision-making.

Lessons learned

- Do not ignore the contribution popular media images and portrayals of an animal might have in helping to drive conflict (or the potential role they might play in transforming conflict).
- Conflict is not only found between stakeholder groups. It is important to look for conflict that might be occurring within groups that look homogenous at first.
- In some cases, opposing sides of a conflict all believe that they have the best interest of the wildlife at heart. This can be common ground to build upon.

—*The Editors*

References

Anderson, M. V. and Henderson, A. J. (2005). Pernicious portrayals: the impact of children's attachment to animals of fiction on animals of fact. *Society and Animals*, **13**(4):297–310.

Bejder, L., Samuels, A., Whitehead, H., et al. (2006). Decline in relative abundance of bottlenose dolphins exposed to long-term disturbance. *Conservation Biology*, **20**(6):1791–8.

Bekoff, M. (2002). *Minding Animals: Awareness, Emotions, and Heart*. New York: Oxford University Press.

Besio, K., Johnston, L., and Longhurst, R. (2008). Sexy beasts and devoted mums: narrating nature through dolphin tourism. *Environment and Planning*, **40**(5):1219–34.

Bryld, M. and Lykke, N. (2000). *Cosmodolphins: Feminist Cultural Studies of Technology, Animals and the Sacred*. New York: Zen Books.

Bulbeck, C. (2005). *Facing the Wild: Ecotourism, Conservation and Animal Encounters*. London: Earthscan Publications Ltd.

Burnett, D. G. (2010). "A mind in the water: the dolphin as our beast of burden." *Orion Magazine*, May/June Issue, <http://www.princeton.edu/history/people/data/d/dburnett/profile/dgbpdfs/BurnettDG_AMind_Orion_2010.pdf>, accessed November 12, 2011.

Christiansen, F., Lusseau, D., Stensland, E., and Berggren, P. (2010). Effects of tourist boats on the behavior of Indo-Pacific bottlenose dolphins off the south coast of Zanzibar. *Endangered Species Research*, **11**(1):91–9.

Cloke, P. and Perkins, H. (2005). Cetacean performance and tourism in Kaikoura, New Zealand. *Environment and Planning D: Society and Space*, **23**(6):903–24.

Cochrane, A. and Callen, K. (1992). *Dolphins and Their Power To Heal*. Rochester, VT: Healing Arts Press.

Constantine, R. (2001). Increased avoidance of swimmers by wild bottlenose dolphins *(Tursiops truncatus)* due to long-term exposure to swim-with-dolphin tourism. *Marine Mammal Science*, **17**(4): 689–702.

Courbis, S. and Timmel, G. (2009). Effect of vessels and swimmers on behavior of Hawaiian spinner dolphins in Kealakekua, Hōnaunau, and Kauhako bays, Hawaii. *Marine Mammal Science*, **25**(2):430–40.

Cressey, J. (2009). *Deep Voices: The Wisdom of Whales and Dolphin Tales.* Victoria, BC: Trafford Publishing.

Danil, K., Maldini, D., and Marten, K. (2005). Patterns of use of Makua Beach, Oahu, Hawaii by spinner dolphins (*Stenella longirostris*) and potential effects of swimmers on their behavior. *Aquatic Mammals*, **31**(4):403–12.

Davis, S. (1997). *Spectacular Nature: Corporate Culture and the Sea World Experience.* Los Angeles, CA: University of California Press.

Delfour, F. (2007). Hawaiian spinner dolphins and the growing dolphin watching activity in Oahu. *Journal of the Marine Biological Association of the United Kingdom*, **87**(1):109–12.

Desmond, J. (1999). *Staging Tourism: Bodies on Display from Waikiki to Sea World.* Chicago, IL: University of Chicago Press.

Donaldson, R., Finn, H., Bejder, L., Lusseau, D., and Calver, M. (2012). The social side of human–wildlife interaction: wildlife can learn harmful behavior from each other. *Animal Conservation*, **15**(5):427–35.

Evernden, N. (1999). *The Natural Alien: Humankind and Environment*, 2nd edn. Toronto, ON: University of Toronto Press.

Evernden, N. (2003). Nature in industrial society. In Armstrong, S. J. and Botzler, R. G. (eds), *Environmental Ethics*. New York: McGraw-Hill, pp. 191–9.

Fawcett, L. (1989). Anthropomorphism: in the web of culture. *UnderCurrents*, **1**:14–20.

Fraser, J., Reiss, D., Boyle, P., et al. (2006). Dolphins in popular literature and media. *Society and Animals*, **14**(4):321–49.

Fudge, E. (2002). *Animal.* London: Reaktion Books.

Griffin, D. (1986). Foreword. In Schusterman, R., Thomas, J., and Wood, F. (eds), *Dolphin Cognition and Behavior: A Comparative Approach*. Hillsdale, NJ: Lawrence Erlbaum Associates Publishers, pp. xi–xiii.

Hall, S. E. (2004). *The Legend of Katama: The Creation Story of Dolphins.* Oaks, PA: Island Moon Press.

Herman, L. (2002). Expanding the cognitive world of the dolphin. In Bekoff, M., Allen, C., and Burghardt, G. (eds), *The Cognitive Animal*. Cambridge, MA: MIT Press, pp. 275–83.

Info Grafik. (2014). *Ancient Hawaii.* <www.hawaiihistory.org>, accessed November 25, 2014.

Lilly, J. C. (1961). *Man and Dolphin.* Garden City, NY: Doubleday and Company Inc.

Livingston, J. (1994). *Rouge Primate.* Toronto, ON: Key Porter Books.

Madden, F. and McQuinn, B. (2014). Conservation's blind spot: the case for conflict transformation in wildlife conservation. *Biological Conservation*, **178**:97–106.

McIntyre, J. (1974). *Mind in the Waters: A Book to Celebrate the Consciousness of Whales and Dolphins.* Toronto, ON: McClelland & Stewart.

Montagu, A. and Lilly, J. C.s (1963). *The Dolphin in History: Papers Delivered at a Symposium at the Clark Library October 1962.* Los Angeles, CA: University of California.

Montgomery, S. (2003). Dance of the dolphin. In Frohoff, T. and Peterson, B. (eds), *Between Species: Celebrating the Dolphin–Human Bond*. San Francisco, CA: Sierra Club Books, pp. 110–23.

Newsome, D., Dowling, R., and Moore, S. (2005). *Wildlife Tourism.* Cleveland, UK: Channel View Publications.

Nielsen, A. (1994). *Dolphin Tribe: Remembering the Human Dolphin Connection*, 2nd edn. Kihei, HI: Dancing Dolphin Press.

Nollman, J. (1985). *Dolphin Dreamtime.* London: Anthony Blond.

Norris, K., Würsig, B., Wells, R. S., and Würsig, M., eds. (1994). *The Hawaiian Spinner Dolphin.* Berkeley, CA: University of California Press.

Ocean, J. (1989). *Dolphin Connection: Interdimensional Ways of Living.* Kailua, HI: Dolphin Connection.

Ocean, J. (1997). *Dolphins into the Future.* Kailua, HI: Dolphin Connection.

Östman-Lind, J., Driscoll-Lind, A., and Rickards, S. (2004). *Delphinid Abundance, Distribution and Habitat Use off the Western Coast of the Island of Hawai'i.* Southwest Fisheries Science Center Administrative Report LJ-04-02C. San Diego, CA: Southwest Fisheries Science Center.

Peace, A. (2005). Loving leviathan: the discourse of whale-watching in Australian ecotourism. In Knight, J. (ed.), *Animals in Person: Cultural Perspectives on Human–Animal Intimacies*. New York: Berg, pp. 191–210.

Sacks, D. (1995). *A Dictionary of the Ancient Greek World*. New York: Oxford University Press.

Sandoz-Merrill, B. (1999). *In the Presence of High Beings: What Dolphins Want You To Know*. Tulsa, OK: Council Oak Books.

Santa Barbara Museum of Natural History. (1991). *The Chumash People: Materials for Teachers and Students*. Santa Barbara, CA: Santa Barbara Museum of Natural History.

Selke, I. (1997). *Journey to the Center of Creation: Entering the World of Dolphins and the Dimensions of Dreamtime*. Stanwood, WA: Living From Vision.

Servais, V. (2005). Enacting dolphins: an analysis of human–dolphin encounters. In Knight, J. (ed.), *Animals in Person: Cultural Perspectives on Human–Animal Intimacies*. New York: Berg, pp. 211–29.

Silva, L. (2013). How ecotourism works at the community-level: the case of whale-watching in the Azores. *Current Issues in Tourism*, **18**(3):1–12.

Simmonds, M. P. (2008). The brains of whales. In Armstrong, S. and Botzler, R. (eds), *Animal Ethics Reader*, 2nd edn. New York: Routledge, pp. 193–203.

Slater, C. (2004). *Dance of the Dolphin: Transformation and Disenchantment in the Amazonian Imagination*. Chicago, IL: University of Chicago Press.

Taylor, S. (2003). *Souls in the Sea: Dolphins, Whales, and Human Destiny*. Berkeley, CA: North Atlantic Books.

Tyne, J. A., Pollock, K. H., Johnston, D. W., and Bejder, L. (2014). Abundance and survival rates of the Hawai'i Island associated spinner dolphin (*Stenella longirostris*) stock. *PLoS ONE*, **9**(1): e86132.

Warren, J. (2009). What it's like to be a whale. In Farr, M. and Pearson, I. (eds), *Cabin Fever*. Markham, ON: Thomas Allen Publishers, pp. 303–32.

White, T. (2007). *In Defense of Dolphins: The New Moral Frontier*. Malden, MA: Blackwell Publishing.

Wiener, C. S. (2013). Friendly or dangerous waters? Understanding dolphin swim tourism encounters. *Annals of Leisure Research*, **16**(11):55–71.

Wiener, C. S. (2014). *Friendly or Dangerous Waters? Dynamics of a Natural and Human System in the Coastal Waters of Hawai'i: Understanding Spinner Dolphin Marine Tourism and Human Perceptions*. Ph.D. thesis. Toronto, ON: York University.

Wilson, E. (1993). Biophilia and the conservation ethic. In Wilson, E. and Kellert, S. (eds), *The Biophilia Hypothesis*. Washington, DC: Island Press, pp. 31–41.

Würsig, B. (1996). Swim-with-dolphin activities in nature: weighing the pros and cons. *Whalewatcher*, **30**(1):11–15.

9

Examining Identity-Level Conflict
The Role of Religion

Julie-Beth McCarthy

9.1 Introduction

Identity is embedded within the narratives by which we lead our lives. These narratives shape who we are and how we see ourselves, tell us where we come from, situate us among those around us, and point toward future possibilities (Lederach, 2003). Identity will often dictate the form of conflict, causing people to feel the need to protect their sense of self as well as that of a group (Chapter 1; Lederach, 2003; Madden and McQuinn, 2014). Thus, when a conflict can be interpreted as an attack against a people, a culture, or, for example, a religion, it is an identity-level conflict (Lederach, 2003; Madden and McQuinn, 2014). As noted in Chapter 1, such conflict can be rooted in perceptions of power, historical dynamics, the needs of a group or an individual, local beliefs, and potential prejudices, as well as desires regarding dignity, respect, autonomy, and recognition (Madden and McQuinn, 2014).

Inherent to identity are values. For many people, religion shapes identity by providing a value-based framework that can hold old narratives and assist in the construction of new ones. It situates the person or group in a time and place through the provision of a view of the world and a space within it, along with various practices and codes of conduct that serve to ensure that the individual or group is identifiable as an active participant in the narrative.

Religion is a confluence of its environment: political, cultural, historical, ecological, and economic. Thus, the ecological is not simply the background upon which religion acts but is a contributing factor in its expression (Taylor, 2007). This kind of interaction, while ensuring complexity, creates for the astute conservationist a wonderful opportunity to learn more about the intricate dynamics at play in a given human–wildlife conflict scenario.

9.1.1 Religion and conservation

As White (1967, p. 1205) so eloquently states, "What people do about their ecology depends on what they think about themselves in relation to things around them. Human ecology is deeply conditioned by beliefs about nature and destiny—that is, by religion." Bhagwat,

Human–Wildlife Conflict: Complexity in the Marine Environment. Edited by Megan M. Draheim, Francine Madden, Julie-Beth McCarthy, and E. C. M. Parsons © Oxford University Press 2015. Published 2015 by Oxford University Press.

Dudley, and Harrop (2011) note that more than four billion people (nearly two-thirds of the world's population) living in Conservation International's hotspots follow an organized religion. Indeed, in 2010 approximately 84% of the global population identified with a religious group (Pew Forum on Religion and Public Life, 2012). The importance of addressing conservation issues is being increasingly recognized by religious groups while, simultaneously, conservation organizations are recognizing that religions have much to contribute to the implementation of conservation. While not all groups, or their adherents, view the natural world as sacred, the large percentage of the population that identifies as religious points to an opportunity to seek connections at a deeper level, that of identity.

Nature may be viewed as sacred, and time spent in nature—or even visual and textual representations of it—can evoke a sense of the religious (Jenkins and Chapple, 2011). Many cultures, both mainstream and traditional, venerate nature and grant it intrinsic value, viewing it as a manifestation of the sacred unseen (Mallarach and Papayannis, 2010). In fact, reverence for nature can be interpreted as potential for common ground between conservation and religion: both are driven by a strong sense of ethics. Consequently, faith groups may have much to contribute to conservation efforts, including instances of human–wildlife conflict. These contributions have been categorized in two broad ways, as demonstrated in Figure 9.1: influence, and sacred natural sites (SNS; Dudley et al., 2005).

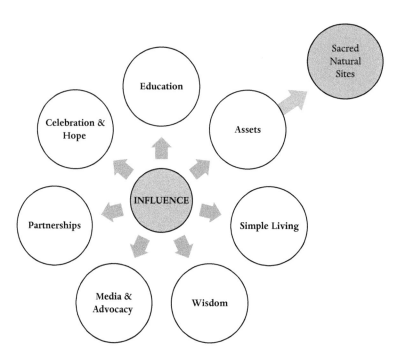

Figure 9.1 Ways in which religious groups influence conservation (adapted from Palmer and Finley, 2003; Finley, 2009).

9.1.1.1 *Influence*

Religion can exert a profound influence on large groups of people through education, assets (investments, land ownership), proscriptions for living, wisdom, advocacy, partnerships, and positive attitude (Palmer and Finley, 2003; Dudley et al., 2005). Religions, though they may at times seem quite static, actually exist in a state of constant change and redefinition (Palmer and Finley, 2003; Taylor, 2007). The construction, deconstruction, and ongoing transformation of a religious narrative comprise a highly creative exercise in adaptability and flexibility.

This ongoing adaptation can open religion to change (Fredericks and O'Brien, 2011). Such change may occur by re-examining texts, practices, and foundational beliefs to identify and support a new path forward (Palmer and Finley, 2003). The wisdom of their teachings are often amicable to ideas of living simply and in harmony with nature, looking to the future/thinking long term, having patience and tolerance, developing trusting relationships, and often most importantly, pronouncing a message of hope and celebration (Palmer and Finley, 2003; Finley, 2009). Ideas like this can support such conservation concepts as the precautionary principle, aide against the pitfalls of shifting baselines, encourage multigenerational timelines and frameworks, promote tolerance in conflict, work within the ebb and flow of wicked dynamics, and bring a positive outlook to a difficult situation. Additionally, religions are able to draw on their substantial networks and media savvy to reach large groups of people, providing educational opportunities and promotion of issues (Palmer and Finley, 2003).

However, it must be noted that faith statements that champion religious environmentalism should not be assumed to be taken up by every group or adherent; the significance of such statements will depend on their lived experiences of nature and their subsequent perceptions of how their beliefs fit within that (Taylor, 2007; Dudley et al., 2009; Jenkins and Chapple, 2011). In the same way, not every religious group will express concern for the environment or be as flexible in their willingness to change and adapt to new ideas. Charges have been leveled against the main monotheistic faiths (Christianity, Islam, and Judaism) for having had a profoundly negative impact on the environment and our view of it, mostly due to ideas regarding dominion over nature (e.g., see White, 1967; for an overview, see Dudley et al., 2005, and Jenkins and Chapple, 2011). While such assertions have been met with faith statements on the importance of environmental protection, ideas of dominion and a resulting unwillingness to change can still be seen in some (often fundamental) elements of many religions (Bhagwat et al., 2011).

The influence of religion can have both positive and negative impacts on a given conservation program and thus also be a key component of conflict during implementation. It is important to be aware of these possible influences. The shared challenge of conservation presents an opportunity to explore shared perspectives and resources that can support empowerment and collaboration. This is important not just between conservationists and religious groups but also among stakeholders with similar religious backgrounds.

Table 9.1 List of possible sacred species and sacred natural sites (adapted from Dudley et al., 2005)

Sacred species	Individual plant
	Individual animal
	Plant species
	Animal species
Sacred natural sites	Enduring/geographic feature
	Natural site
	Semi-natural site
	Built monument within a protected landscape
	Built monument with incidental ecological value
	Landscape/seascape

9.1.1.2. SNS

Religious organizations are among the largest landowners and investors in the world (Palmer and Finley, 2003; Dudley et al., 2005). Land that has been marked sacred by a faith group (owned or not) can greatly contribute to conservation. Such SNSs represent the oldest form of protected area, are found in all ecosystems (including maritime), and cover all IUCN categories (Dudley et al., 2005; Mallarach and Papayannis, 2010). Included within this discussion is the recognition of sacred species. Both SNSs and sacred species manifest in a number of different forms, as outlined in Table 9.1.

As an aspect of traditional natural resource management that is quite common and serves both people and ecology, SNSs and sacred species provide a range of ecosystem services (Millennium Ecosystem Assessment, 2005; Rutte, 2011). While generally protected for their spiritual and cultural value, they may also have value socially, economically, and/or ecologically. Sacredness of a site or a species may or may not imply full protection (Dudley et al., 2005). While many SNSs and sacred species have a long history with a faith, this may not always be the case (Mallarach and Papyannis, 2010). Just as faith groups may revisit old, established teachings and adapt them to new situations, they may deem a new site or species to be sacred based on new needs and experiences. As a result, conflict that is connected to SNSs may manifest in a number of different ways and with varying intensity over time (Rutte, 2011).

9.1.2 Traditional ecological knowledge and taboos

Berkes (2008, p. 7) defines traditional ecological knowledge (TEK) as "a cumulative body of knowledge, practice, and belief, evolving by adaptive processes and handed down through generations by cultural transmission, about the relationship of living beings (including humans) with one another and with their environment." According to Berkes (2008), TEK consists of four interconnected components: local knowledge of the environment, systems of resource management, social institutions, and worldview (the spiritual and cultural aspects of TEK). Utilizing this definition, we can see that applying the concept of religion to TEK can be misleading, as it attempts to isolate one component of a holistic framework, whereas "religion" may be inseparable and unintelligible outside of ecological and social dynamics (Jenkins and Chapple, 2011). For our purposes, it may be helpful to think of religion as providing the worldview and belief structure within which TEK and its resulting practices manifest.

Taboos are an expression of TEK. They are social norms that can provide functional management of sites and species by dictating when a resource can be accessed in time (temporal taboo), which members of society can access a resource (segment taboo), how a resource can be accessed (method taboo), where a resource may be accessed (habitat taboo), what species can be accessed (species-specific taboo), and at what life stage (life-history taboo; Colding and Folke, 2001). It can be quite difficult to pull apart the origins and functions of a taboo—be it ecological, social, or religious—and a taboo can either intentionally or unintentionally have protective value for a site or species (Colding and Folke, 2001). Oftentimes SNSs and sacred species are either partially or fully protected via taboos rooted in religious belief (Colding and Folke, 1997; Rutte, 2011). As in the case of SNSs and sacred species, taboos need not be old or well established. They can be reinvented, reinvigorated, or newly generated once new information and experiences have been imparted to stakeholders (Cohen and Foale, 2011).

9.1.3 Religion and human–wildlife conflict

The interplay between conservation and religion is complex, and indeed no simple relationship exists for its articulation. Subsequently, the ways in which religion may impact human–wildlife conflict cannot be broken down into a step-by-step formula. Religion, "as an aspect of culture [that is] both living and lived" (Taylor, 2007, p. 134), can profoundly shape the outlook of a stakeholder, and through his/her behaviors and attitudes, that stakeholder can considerably impact the shape of a conflict by determining the extent of his/her reaction and level of tolerance (Madden, 2004).

One should not assume that, as a result of this complexity, religious dynamics will work against a process. Oftentimes religion can be a valuable ally, causing unforeseen opportunities to arise (e.g., see Box 9.1) and enabling the rebuilding of tolerance that has eroded over time (Madden, 2004). Teachings, texts, practices, and beliefs can be revisited and built upon. One can seek out connections across stakeholders by looking for the religious dimensions in conservation practices (Jenkins and Chapple, 2011) and draw on a community's religious narratives as "teaching aids" for understanding local values (Madden, 2004).

A religious community will have a unique set of values and needs associated with a particular conservation process. It is worth noting that the religious rewards a community may derive from an SNS or a sacred species can be highly valuable to a conservation process and

---

Box 9.1 Report from the field—El Santo Niño of the coral reef

I heard a profoundly important yet humorous anecdote at the workshop we lead in Cebu, Philippines. Project Seahorse gathered colleagues from many sectors for a discussion around our research findings on marine protected areas (MPAs) under the mantra of "MPAs in the Philippines: Ever more, ever better."

We were delighted that some Roman Catholic priests from Sea Knights participated in the workshop. This is a group of ocean-loving priests; many of them are scuba divers. One of the Sea Knights, Father Tito, is also the Executive Director of the Santo Niño de Cebu Augustinian Social Development Foundation.

Father Tito told us this wonderful story of how the Church took a very sacred statue of the Santo Niño (the infant Jesus) from the basilica in Cebu to islands on Danajon Bank, the area of the Philippines where we have done so much of our MPA research and management. The arrival of the Santo Niño created a very festive occasion, with huge village gatherings, boat parades, and street processions.

During the celebrations, the Church showed a film on environmental responsibility, motivated by its desire for stewardship of God's creation. A large international fisheries management aid project on the reef later reported a notable decrease in poaching rates in the villages that the icon visited.

Thereafter, the municipal government of Bien Undo, Bohol placed two replica icons of the Santo Niño and the Blessed Virgin Mary underwater in an area with considerable illegal fishing, primarily using dynamite. The mayor of this municipality took the initiative because he is himself a Sea Knight.

Apparently the fisheries abuse disappeared because, after all, you do not blow up the Infant Jesus, or His Mother. Such a change is especially important because both icons lie in a critically important newly established MPA full of corals and fish that are vulnerable to dynamite fishing.

Dr. Amanda Vincent, June 27, 2011 (Vincent, 2011)

---

are no less valid than the analogous values and needs of, say, an ecotourist (Madden, 2004). A lack of awareness on the part of a conservationist can mean a failure to meet these needs, thus significantly impacting outcome. In an examination of the establishment of two national marine areas, Fiske (1992) highlights the fact that the success of a project can be dependent on whether local cultures and values are included or ignored. And indeed, this is an observation that deeply resounds within the field of human–wildlife conflict (Madden, 2004; Madden and McQuinn, 2014).

Part of ensuring the inclusion of local values is recognizing that their worldview may include a different "place-view" (Higgins-Zogib, 2007). This may refer to an SNS or may simply be an expression of how a stakeholder views their local environment—either way, it is a site-specific manifestation of the worldview. Failure to properly engage with this place-view can result in conflict; however, if knowledge of this perspective is well applied, both biodiversity and cultural diversity can benefit (Higgins-Zogib, 2007). It is important to recognize that the place-view of a stakeholder may be quite different from that of the conservationist, though both will be looking at the same space. Working to bridge the gap between these potentially disparate place-views is often an integral component of a conflict transformation process. One of the principles of conflict transformation is that it is important to consider place, position, presence; and part of understanding this principle is acknowledging that where people meet is important (Madden, personal communication). Consequently, in some situations, meeting at a place that has both religious/sacred meaning and conservation meaning might be a way to bring two sides together (Madden, personal communication).

Higgins-Zogib (2007) points out that the balance between the place-views of stakeholders when an SNS is present can be disrupted when there is no common understanding of a sacred place, when there is no respect for the sacred elements of a space, and/or when the effects of one place-view are damaging to the objectives of another. These issues point to areas of potential conflict and are worth consideration. Of course, these issues are not unique to religious values and could easily be a general list of reasons why people get into conflict over a particular space or species: lack of understanding of another's values related to a space/species, lack of respect for particular values, and contention over how a space/species should be protected are all common catalysts for human–wildlife conflict.

Rutte (2011) highlights some further potential points of conflict that integrate and expand upon these components (see Table 9.2). These points of conflict will be further discussed in Section 9.3.1.

Table 9.2 Potential conflicts related to sacred natural sites and species

Point of conflict		Example
Competition over natural resource use		Along the Tanzanian coast, local worldviews lack a linear concept of time and causation, resulting in fishing practices that do not take into consideration future abundances (Palmer and Finley, 2003; Dudley et al., 2005).
Tradition vs. modernization	Spiritual vs. economic values	The Saintes-Maries-de-la-Mer sacred natural site in the National Regional Park of the Camargue is the site of three annual pilgrimages; however, the spiritual, cultural, and natural heritage of the site is threatened by economic demands related to tourism and urban expansion by the local municipality. Management of the park has not integrated such religious values despite the fact that it would broaden support for the conservation goals (Mallarach, 2011).
	Spiritual vs. ecological values	In Mexico, sea turtles are considered fish and thus seen as an appropriate meal for Catholics to consume during Lent and on Fridays, when meat has traditionally been taboo. Conservation organizations have sought to have sea turtles declared as meat by authorities within the Church in an effort to reduce illegal harvesting of the endangered species (Nichols and Palmer, 2006).
Changing spiritual values		Along Kenya's southern coastline, animistic indigenous traditions are slowly being assimilated by the spread of Islam. Kaya, which are former burial sites within the coastal forests, are no longer used as burial sites but are still revered as sacred land associated with ancestors. Care for these sites is starting to lag due to shifts in beliefs. Thus, these sites are undergoing their third change in spiritual value (McClanahan et al., 1997).

9.2 Case studies

An examination of two case studies highlights some of the above-mentioned concepts. The first case study provides an example of a problem that has yet to be adequately addressed, in large part due to the disparate worldviews involved. In this case study, the local population holds a worldview that, on the surface, does not align with current conservation management approaches. Applying a levels of conflict approach to this case provides some insight in regards to the many and complex dynamics that need to be addressed for conservation success. The second case provides an example of a successful campaign that focused on the role of identity in preventing conflict and increasing likelihood of success. In this case study, the campaign managers included social and cultural demographics in their initial planning and adopted an approach that not only integrated these components but put them at the fore-front of the campaign. This rooted the endangered species directly into the local belief structure, thus strengthening and deepening the community's connection to the species.

9.2.1 Conflicting worldviews: the Bajau

9.2.1.1 Background

The Bajau are one of three distinct sets of "sea nomads" living among the archipelagic environments of insular Southeast Asia (Magannon, 1998). They have traditionally lived a wholly nomadic lifestyle consisting of maritime foraging over an expansive area (following fishing conditions, kin obligations, and political situations) but are now mostly settled in a few coastal "floating" villages (Crabbe, 2006; Clifton and Majors, 2012). They are a socially, culturally, economically, and politically marginalized ethnic group (Magannon, 1998; Shepherd and Terry, 2004; Clifton and Majors, 2012).

The Bajau are simultaneously recognized for their extensive knowledge of the marine environment and considered a major threat to conservation. A division exists between those who see them as central to the success of local conservation efforts and those that claim the Bajau lack the long-term commitment necessary for meaningful conservation (Clifton and Majors, 2012; Karam et al., 2012). This is mainly due to the fact that the Bajau's worldview exists in stark contrast to contemporary marine conservation practices. Conflict arising out of conservation processes will require an astute understanding of each of the levels of conflict. Of particular difficulty is the identity-level conflict, which requires a sound understanding of the Bajau worldview and an extensive reconciliation process if it is to be properly addressed.

9.2.1.2 The dispute

In 1996, Wakatobi Marine National Park (MNP), located in the Tukangbesi archipelago (see Figures 9.2 and 9.3), was created with the express aim to protect coral reefs and fish stocks while promoting ecotourism as an alternative income source for local fisherman (Shepherd and Terry, 2004). There are five Bajau villages nestled within the MNP, and the MNP's strategies for marine protection target the inclusion of the Bajau in management (Shepherd and Terry, 2004; Stacey, 2007). However, destructive fishing techniques practiced by the Bajau,

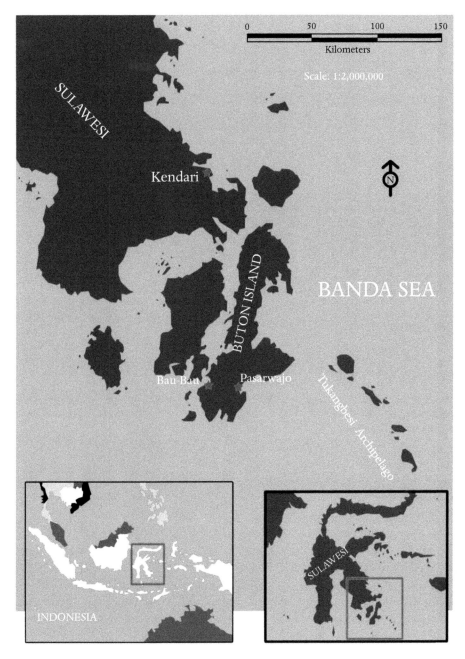

Figure 9.2 Southeast Sulawesi and the Tukangbesi Archipelago. The islands shown in Figure 9.3 are in the archipelago southeast of the island of Sulawesi (Crabbe, 2006, p. 58).

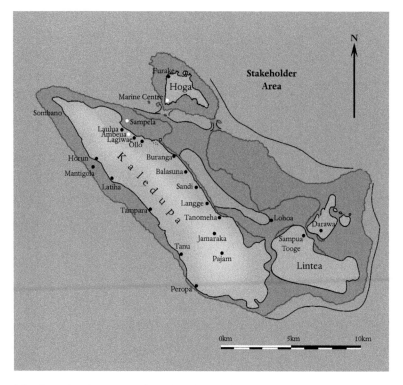

Figure 9.3 Islands and settlements in the Tukangbesi archipelago, south east of Sulawesi in Indonesia (see Figure 9.2). The Bajau community at Sampela is situated between the islands of Hoga and Kaledupa in the Wakatobi Marine National Park (Crabbe, 2006, p. 58).

including bomb and cyanide fishing, paired with coral mining, have damaged large parts of the local reef systems (Shepherd and Terry, 2004; Crabbe, 2006).

The destructive fishing techniques employed by the Bajau and the coral mining utilized to lay the foundations for their "floating houses" are cause for dispute within the MNP (see Figure 9.4). Furthermore, the absence of a linear concept of time and temporal causation, seen in the perception that daily fish catches are not affected by past effort but rather are the result of current effort and the grace of the sea spirits, have clear implications for local conservation goals (Clifton and Majors, 2012). Despite the fact that people are usually more inclined to protect their natural environment when it is so intricately linked to their worldview (Higgins-Zogib, 2007), the absence of future consequences in the Bajau has managers and academics debating whether they can actually be effectively integrated into management (Karam et al., 2012).

9.2.1.3 Underlying and identity-level conflict

Madden and McQuinn (2014, p. 102) state that "identity-based conflicts find their expression in the relationship among communities, between a community and conservation authorities

Figure 9.4 The Bajau village of Sampela, which is built on top of corals from local reefs that are mined by the villagers (photo by M. James C. Crabbe) (Crabbe, 2006, p. 59).

or the state, or even between conservation groups competing with one another toward the same conservation goals." Such relationships are fraught with complex power dynamics, and identity is closely linked to power; as a result it is very important in a conservation conflict situation to be aware of how people perceive this link (Lederach, 2003). It is especially important to be aware of the interplay of identity and power when dealing with a community who feel that their identity is under threat or being eroded, or who have an ongoing history of marginalization (Lederach, 2003), like that of the Bajau.

The Bajau have been marginalized from society for centuries. Ongoing political development programs showcase the power of the state over this small group, with the MNP itself being a further expression of this power dynamic. Protected areas that are established and reliant upon a state or similar central authority are more likely to generate conflict related to the SNSs and sacred species within its bounds (Higgins-Zogib, 2007). People feel they have lost control of what is sacred to them, and additional stressors such as increased tourism and the lack of knowledge of most hired managers work to compound communities' feelings of powerlessness (Higgins-Zogib, 2007). It is likely that the Bajau harbor such feelings with regards to the MNP, while the MNP managers doubt the Bajau's ability to be active participants in MNP management.

In cases where indigenous knowledge must be integrated into planning and management, the interplay of power and identity is highlighted not only because of concerns of cultural disintegration but also because these power dynamics commonly manifest when scientific knowledge and TEK compete (Fredericks and O'Brien, 2011). As noted in Chapter 1 and in

Madden and McQuinn (2014), the process in which planning and management is undertaken can be a significant source of conservation conflict (Madden and McQuinn, 2014). Being primarily concerned with the exercising of decision-making authority, said processes can exacerbate conflict by playing into perceptions of power and may have unintended consequences (Madden and McQuinn, 2014). For example, failure to account for the Bajau worldview may result in members of the community being unwilling or unable to answer questions posed in surveys and interviews framed in linear (i.e., foreign) concepts of time and space (Clifton and Majors, 2012). The result of this can be the establishment of regulations that cause confusion, further adding to local mistrust of authority, and the continuance of the Bajau fishermen's lack of support (Clifton and Majors, 2012). For the Bajau, the benefits of destructive fishing techniques have been absorbed into existing community practices, with the result that these fishermen gain improved social status through catch sharing (Clifton and Majors, 2012). This is a potent combination, as identity becomes linked to destructive habits and spiritual values become enmeshed in the economic values.

The Bajau worldview is a rich belief system populated by a multitude of spirits, and relationships with these spirits determine local taboos (Magannon, 1998; Stacey, 2007; Karam et al., 2012). This worldview reflects tidal movements and seasonal changes, causing the passage of time for the Bajau to not fit within a linear model (Clifton and Majors, 2012). Sea spirits are central to the Bajau worldview and can reside in any number of locations, permanent or transitory (e.g., in rocks, waves, coral reefs, tidal currents, etc. (Clifton and Majors, 2012)). The main function of these spirits is to reward and punish fishermen; rewards include fishing success while punishment can be seen as mishaps at sea, illnesses, and conflicts (Magannon, 1998; Clifton and Majors, 2012). Temporary fishing success is a result of *padalleang*, or good fortune from a spirit, and is not related to fish abundance, fishing technique, individual skill, or any other modern management understanding (Clifton and Majors, 2012).

The worldview of the Bajau reflects a notion of sacred nature wherein values and beliefs are fully integrated into their environment, underscoring Berkes' definition of TEK. This case highlights the fact that many threatened areas and species exist in places where the local people possess a worldview that is in stark contrast to those of Western science. A divergent worldview such as that of the Bajau can contain different understandings of time and space, including the lack of a concept of "future," which can easily lead to confusion, confrontation, misunderstandings, and misguided applications of management (Crabbe, 2006). Further complicating the issue is that there is no text to learn from. Integration into the community and/or the trust and guidance of a local spiritual leader are long-term but necessary commitments to gain a functional understanding of local values.

Additionally, Bajau beliefs are syncretic, meaning that they have absorbed elements of a separate, often contradictory belief system: in this case, Islam (Sunni; Stacey, 2007). These elements are most generally found in day-to-day activities and are observed with varying degrees by each individual (Stacey, 2007). This syncretism probably dates back as far as the 1400s, when the Bajau took to sea to avoid conversion to Islam (Magannon, 1998). Political actions dating back to the 1950s (Stacey, 2007) have continued this integration, which persists today through current development schemes of the Indonesian government (Karam et al.,

2012). Clearly, there exists a long history of conflict between the Bajau and "others"; this history is a defining characteristic for the community (Magannon, 1998). Consequently, ongoing marginalization is an aspect of underlying conflict that has become so ingrained that it is also a defining feature of Bajau identity (see Comment).

Comment

For more on how a history of oppression and colonialism can influence present-day conflict, see Chapter 4 for Christine Gleason's thoughts on how a history of colonialism might play a role in determining the effectiveness (or lack thereof) of a co-management system for whale-watching in the Dominican Republic.

—The Editors

The underlying history of conflict between the Bajau and the government, paired with the Bajau's nonlinear concept of time and their perception of the role of sea spirits in determining fishing success, creates a challenging situation for ensuring protection of the MNP.

9.2.1.4 *Discussion*

The debate as to whether the Bajau can effectively aide in local management highlights the difficulty in working with conflict that is strongly rooted in identity. This is a conflict scenario that can be hard to fix; however, if work can begin to reconcile these differences, it can lead to better outcomes, ensuring that fixes are not superficial in nature (Madden, 2004). Spiritual values are also in conflict with ecological values in Madagascar, where the Vezo, a semi-nomadic ethnic group, define themselves in part through the hunting and consumption of sea-turtle meat. Like the Bajau, the catch is distributed within the community, and such sharing is a respected ritual linked to spirit appeasement (Lilette, 2006). Despite legislation prohibiting the hunting and sale of sea-turtle parts, both persist, as enforcement is minimal, enabling the traditions to continue to provide spiritual benefits as well as economic stability (Lilette, 2006). The few effective prohibitions with the Vezo have been those rooted in local taboos (*faly*; Lilette, 2006). Prohibitions that derive from existing local values often are successful, as they develop in a context that is meaningful to the community and can become part of local identity. A similar example exists from Masili on the Tanzanian coast, where textual references of the local Islamic religion were drawn on to promote less destructive, sustainable fishing practices (Palmer and Finley, 2003; Dudley et al., 2005). Oral traditions and taboos within the Bajau may exist that could provide similar options for coexistence. Embracing the unique aspects of the Bajau's worldview instead of seeing them as obstacles to success may generate unique opportunities for collaboration. Indeed, within the Bajau cosmology, certain species are subject to taboos. This has led to calls to incorporate these taboos into ecotourism schemes (Karam et al., 2012). A proposal to integrate Bajau knowledge of whale shark behavior (see Karam et al., 2012) is one case that seeks to do just that and shows that collaboration may be possible. Recognizing local beliefs about animals within a human–wildlife conflict process is important (Madden, 2004); such initiatives have the potential to create a bridge of common ground leading to shared understanding from which other programs can grow. Patience and a willingness to work within the natural ebb and flow of relationships are important skills to exercise when worldviews and place-views are quite different. The underlying and identity-level conflict highlighted in this case study is too often ignored or downplayed by conservation and government agencies, when in fact long-term, sustainable conservation success hinges on it being addressed (Madden and McQuinn, 2014).

9.2.2 Finding solutions: whale sharks on the Indian coast

> This marvel of nature and a gift of God, should be the pride of Gujarat.
>
> **Morai Bapu (Chaudhary et al., 2008)**

9.2.2.1 Background

In the 1980s whale shark landings on the Gujarat coast (see Figure 9.5) went from being incidental to being the result of a targeted commercial fishery (Pravin, 2000; Hanfee, 2001). Initially targeted for their liver, the sharks quickly started being harvested for their meat, skin, cartilage, and fins as well (Pravin, 2000; Hanfee, 2001). This shift was driven by declines in traditional fisheries, and increased demand for shark parts in India and parts of Asia (Pravin, 2000; Hanfee, 2001). Upon realizing the large extent of this fishery and having taken note of the whale sharks' threatened status globally, conservationists (Pravin, 2000), TRAFFIC (Hanfee, 2001), and a documentary film entitled *Shores of Silence* (2001) all called for the whale shark to be protected within India and internationally under CITES regulations. On May 28, 2001, the Indian government's Ministry of Environment and Forests added the species to Schedule I of the Wildlife (Protection) Act, 1972, under Subsection (1) of Section 61, thereby granting the sharks full legal protection in Indian territorial waters (Chaudhary et al., 2008). This response was shortly followed by the listing of whale sharks under CITES Appendix II in 2003 (UNEP-World Conservation Monitoring Centre, 2013).

A conflict arose between the state of Gujarat and the fishermen once protection was in place; hunting continued and enforcement had the logistical challenge of a long coastline and limited resources (Chaudhary et al., 2008). To address this conflict and aid in the conservation of the whale shark, a "Save the Whale Shark" campaign was launched in 2004 by Wildlife Trust of India, International Fund for Animal Welfare, TATA Chemicals, and the Gujarat Forest Department (Chaudhary et al., 2008). A summary of the campaign is provided in the next section.

9.2.2.2 The "Save the Whale Shark" campaign

The following summary is adapted from an account by Chaudhary et al. (2008). From the outset, organizers targeted aspects of local identity—promotion of Gujarat pride championed by a prominent religious leader. The pride campaign drew on Gujarat being home to

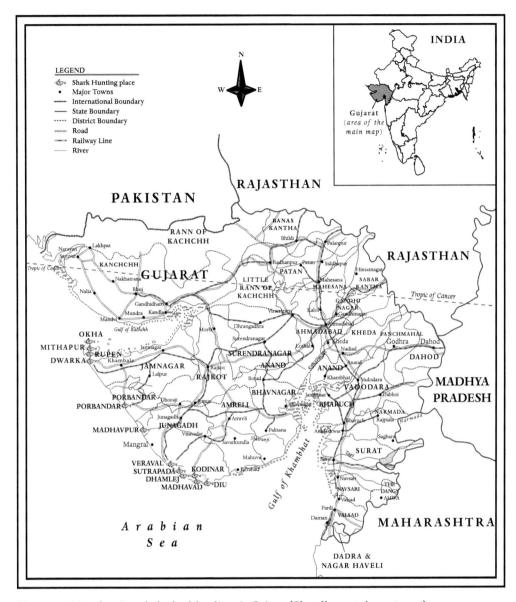

Figure 9.5 Map showing whale shark landings in Gujarat (Chaudhary et al., 2008, p. 16).

the world's largest fish and hinged on the belief that "instilling the sense of conservation in the people on the basis of religious beliefs was a better and a more permanent way of protecting our natural heritage than policing" (Chaudhary et al., 2008, p. 28). It was noted in the preliminary stages that the stretch of coastline lay between two prominent Hindu pilgrimage sites and that in Hindu mythology one of the avatars of Lord Vishnu is a large blue fish closely resembling a whale shark. The coastline communities had been thought to be dominantly Muslim in religious practice; however, surveys and interviews found that the fishing communities actually contained Kolis and Kharwas (Hindu castes). A local religious leader, Morari Bapu, was brought on board to champion the cause. The campaign launched on January 20, 2004, with Morari Bapu taking the lead. He drew on local narratives familiar to the community—equating the whale shark with a religious story, honoring guests as God (the whale shark being a guest to Gujarat), and promoting nonviolence. These messages were promoted throughout the campaign, a play was performed regularly, and a 40 ft. inflatable whale shark provided visual support to the message (see Figure 9.6). The play told the story of a daughter returning to her father's house to give birth. The father sets out to hunt a whale shark to gain some quick cash and things start to go wrong in his relationship with his daughter. The daughter learns of a fisherman arrested for hunting the whale shark and subsequently learns of the sharks' protected status. The daughter, named Vhali ("loved one") gets her father to vow to stop hunting the shark and to protect it as he would his own daughter. From this narrative, the whale shark became known as Vhali; no longer referred to as "barrel" (its local name derived from the method with which the shark is caught). Through the campaign, six communities adopted the whale shark as their city mascot. An auspicious day in the Hindu calendar was declared "Whale Shark Day," to be celebrated annually. Nine

Figure 9.6 The whale shark inflatable model poses for the coastal community at Rupen (Chaudhary et al., 2008, p. 30).

months into the campaign, the tide began to turn, with the news that a fisherman had released an entangled whale shark from his net, citing aspects of the campaign and the sharks' protected status. Due to inquiries and realization of costs, a compensation program was established to further encourage such behavior. More releases followed. The campaign has met with much local success as well as being recognized with awards internationally.

9.2.2.3 Current situation

The whale shark continues to be protected today. Stories of fishermen freeing sharks appear in the media regularly, and between 2010 and 2012 fishermen claim to have released 105 whale sharks (Kaushik, 2013; Times of India, 2013). In 2014 the Forest Department of Junagadh District in Gujarat was granted an award in co-management for the ongoing rescue of whale sharks (DeshGujarat, 2014). Wildlife Trust of India continues to work toward adding to the scientific understanding of whale sharks and their habits along the Gujarat coastline (Wildlife Trust of India, 2011). Ecotourism is being promoted and explored as a viable industry, and Whale Shark Day continues to be promoted and celebrated (Wildlife Trust of India, 2011).

Whale sharks continue to be exploited in other states, most notably in Kerala, Tamil Nadu, and the Godavari regions (Sudhi, 2012; Oppili, 2013; Murali Sankar, 2014). One news piece noted that unlike Gujarat, Tamil Nadu did not have a compensation mechanism in place to encourage releases (Oppili, 2013). However, it is interesting to note that in Gujarat the compensation scheme (a dispute-level solution) was not the driving force behind successful conservation. Instead, it followed and supported an already successful conservation effort. Had it preceded the identity-based campaign, conflict would have focused on compensation, likely resulting in retaliations when compensation payment did not go as expected. The identity-focused campaign enabled the conflict to reach reconciliation as opposed to settlement, reducing the likelihood and magnitude of potential future setbacks (Madden and McQuinn, 2014).

9.2.2.4 Discussion

Hinduism is one of the world's largest faiths. Though areas of India are Muslim, many areas follow various articulations of Hinduism. The foresight of the campaigners to do the necessary preliminary surveys and interviews and to make note of the proximity to local sacred sites instead of just assuming religious affiliation was paramount to the success of this campaign. People are often more inclined to protect a site or species when it is rooted in their belief system (Higgins-Zogib, 2007). Linking whale sharks to the reverential sentiment of the community strengthened the connection to the species. By creating this link, the campaigners brought aspects of identity to the fore and acknowledged their importance to conservation success.

A World Bank study in 1999 (Narayan et al., 1999) found that impoverished populations often trust their local religious institution more than any other. It is fully incorporated into their civil society and provides feelings of ownership, trust, and reliability (Narayan et al., 1999). In this case, the proximity of major sacred sites pointed to religion as a possible bridge builder in this conflict. As previously stated, a sacred site or species need not already be established, but can be created (e.g., see Box 9.1; Mallarach and Papayannis, 2010). Linking the whale shark to the collective identity of both the rural and urban populations increased its value and reinforced the need for its protection.

The language attached to a species can have a profound effect on their perception. Revisiting local religious narratives and adapting them to fit a current need enabled the whale shark to be further integrated into the collective conscious of the community. It also made the ecology of the whale shark relatable to the lives of the people, drawing on familiar tropes such as having a guest visit and the birth of a grandchild. This led to the reframing of the whale shark within the local lexicon. The removal of the negative, harvest-centric name "barrel" (derived from the use of plastic barrels in hunting the sharks) and its replacement with the religious, protection-centric name "Vhali" (loved one) signaled a change in the perception of the species. It was no longer a commodity or resource but a valued member of the community.

Threatened species are regularly harvested for their economic value. The demand for fins described in this case study and an ongoing driver of whale shark landings in other Indian states is a component of one of the most widely known religion-driven conservation conflicts, that of traditional Chinese medicine and culture. This conflict specifically highlights Rutte's modernization-versus-tradition category of conflict, through the spiritual-versus-economic subcategory. In this case, the campaigners reversed this conflict, creating the spiritual to counteract the economic.

9.3 Recommendations

In addressing identity concerns to transform conflict, it is important to structure the process so that people feel they can positively address and articulate their sense of identity (Lederach, 2003). This requires that attention be paid to the language being used, appeals to identity be encouraged, that the process be as dynamic as possible, and that it be guided by honesty, iterative learning, and appropriate exchange (Lederach, 2003). Conservation conflict transformation should support close collaborative work between managers and religious custodians in order to build effective and trusting relationships and grow capacity and knowledge so that shared moral ground and common methods for confronting challenges can be discovered (Madden, 2004; Dudley et al., 2009; Madden and McQuinn, 2014). Further recommendations when working with religious aspects in a conflict include the following:

- Think holistically about all values, and look for improved options for the integration of sacred values in planning by ensuring that cultural impacts and potential cultural factors supportive of coexistence are identified and included in environmental risk assessments (Madden, 2004; Higgins-Zogib, 2007; Dudley et al., 2009).
- Understand sacred values, acknowledge their legitimacy, and emphasize sensitivity, recognition, honesty, and tolerance (Madden, 2004; Higgins-Zogib, 2007; Dudley et al., 2009).
- Recognize that identity is linked to power; be aware of these dynamics and the underlying conflicts that may be present (Lederach, 2003; Madden and McQuinn, 2014).

- Engage in ongoing research and collaboration with social science, working to improve involvement of faith groups in conservation processes. Decision-making processes that work with religion create opportunities for problem-focused investigations into how religious communities interpret specific problems and thus may allow for closer collaborations in the future and at broader management levels.
- People often hold on to their traditional values, even when they are nominally members of a new faith (Dudley et al., 2009; Rutte, 2011), as can be seen in the syncretic nature of Bajau beliefs (see Section 9.3.1). Understand that local values can be rooted in a combination of sources, and resist making assumptions about such roots. Cultural disintegration is a global concern not only in terms of the loss of cultural diversity but also for what it means for the TEK of local ecologies and biodiversity (Higgins-Zogib, 2007; Berkes, 2008).
- Support intercultural dialog, joint education programs, collaborative research, co-planning exercises, and creative activities (Higgins-Zogib, 2007; Mallarach and Papayannis, 2010).
- Don't forget about marginalized spiritualities (e.g., contemporary nature religions) or the diverse experiences within mainstream faiths and indigenous groups (Jenkins and Chapple, 2011), and remember that religion, like people, is dynamic and adaptable; it can bring a positive message of hope and celebration to a process (Palmer and Finley, 2003; Higgins-Zogib, 2007).

9.4 Conclusion

Even where cultural and spiritual beliefs foster tolerance for wildlife, that tolerance is easily eroded when conservation initiatives fail to reflect local value and voices.
(Madden, 2004, p. 240)

The religious viewpoint of a community is enmeshed in that community's culture, which can have a profound influence on how it, or an individual member of the community, will respond to a conflict (Madden, 2004). Consequently, it is imperative that conservationists work toward a holistic understanding of conflict that includes these factors. This can only be accomplished if all parties are willing to overcome past prejudices and trust issues by learning about each other's values, including the language used to communicate them (Mallarach and Papayannis, 2010). This is, however, easier said than done, as people are usually quite invested in the maintenance of their prejudice/distrust (see Chapter 1). Conservation conflict transformation processes can help parties move forward out of such mindsets, and successful integration will not only achieve the aim of preserving nature but will also help to ensure the continuance of local culture (Chapter 1; Madden, 2004; Millennium Ecosystem Assessment, 2005; Mallarach and Papayannis, 2010). The complexity inherent to identity-level conflict can be daunting; however, it is worth addressing such conflict, as the dividends are longer-lasting, more effective relationships.

> **Lessons learned**
>
> - Sometimes it is possible to link the species or resource that people are in conflict over to a community (through culture, religion, history, etc.). This can change the context of the conflict and create an environment of increased collaboration.
> - When culture and worldviews are disparate, find ways to embrace the differences that might lead to opportunities for collaboration.
> - Although compensation programs are often seen as a solution, if they are not preceded or supported by efforts to address underlying and identity-based conflict, they have a good chance of failing.
>
> *—The Editors*

References

Berkes, F., (2008). *Sacred Ecology*, 2nd edn. New York: Taylor & Francis.

Bhagwat, S., Dudley, N., and Harrop, S. R. (2011). Religious following in biodiversity hotspots: challenges and opportunities for conservation and development. *Conservation Letters*, 4(3):234-40.

Chaudhary, R. G., Dhiresh, J., Mookerjee, A., Talwar, V., and Menon, V. (2008). *Turning the Tide: The Campaign To Save Vhali, the Whale Shark (Rhincodon typus) in Gujarat*. Uttar Pradesh: Wildlife Trust of India.

Clifton, J. and Majors, C. (2012). Culture, conservation, and conflict: perspectives on marine protection among the Bajau. *Society and Natural Resources*, 25(7):716-25.

Cohen, P. and Foale, S. (2011). Fishing taboos: securing Pacific fisheries for the future? *SPC Traditional Marine Resource Management and Knowledge Information Bulletin*, 25:3-13.

Colding, J. and C. Folke. (1997). The relations among threatened species, their protection, and taboos. *Conservation Ecology*, 1(1):6.

Colding, J. and Folke, C. (2001). Social taboos: "invisible" systems of local resource management and biological conservation. *Ecological Applications*, 11(2):584-600.

Crabbe, M. J. C. (2006). Challenges for sustainability in cultures where regard for the future may not be present. *Sustainability: Science, Practice, & Policy*, 2(2):57-61.

DeshGujarat. (2014). "National award to efforts in Gujarat that saved 405 whale sharks." *DeshGujarat*, May 28, 2014, <http://deshgujarat.com/2014/05/28/national-award-to-efforts-in-gujarat-that-saved-405-whale-sharks/>, accessed July 12, 2014.

Dudley, N., Higgins-Zogib, L., and Mansourian, S. (2005). *Beyond Belief: Linking Faiths and Protected Areas to Support Biodiversity Conservation*. Gland: WWF International.

Dudley, N., Higgins-Zogib, L., and Mansourian, S. (2009). The links between protected areas, faith, and sacred natural sites. *Conservation Biology*, 23(3):568-77.

Finley, V. (2009). Alliance of Religion and Conservation (ARC). *Contemporary Issues*. Oxford University School of Geography and Environment, unpublished.

Fiske, S. (1992). Sociocultural aspects of establishing marine protected areas. *Ocean and Coastal Management*, 17(1):25-46.

Fredericks, S. and O'Brien, K. (2011). The importance and limits of taking science seriously: data and uncertainty in religion and ecology. In Bauman, W., Bohannon, R., and O'Brien, K. (eds), *Inherited Land: The Changing Grounds of Religion Ecology*. Spokane, WA: Wipf & Stock, pp. 42–63.

Hanfee, F. (2001). *Gentle Giants of the Sea, India's Whale Shark Fishery: A Report on Trade in Whale off the Gujarat Coast*. New Delhi: TRAFFIC-India/WWF-India.

Higgins-Zogib, L. (2007). Sacred sites and protected areas: an interplay of place-views. In COMPAS (Comparing and Supporting Endogenous Development), *Endogenous Development and Bio-Cultural Diversity: The Interplay of Worldviews, Globalization, and Locality*. Geneva, Switzerland, October 3–5, 2006. Geneva: COMPAS.

Jenkins, W. and Chapple, C. K. (2011). Religion and environment. *Annual Review of Environment and Resources*, **36**:441–63.

Karam, J., Meekan, M., Ninef, J., Pickering, S., and Stacey, N. (2012). Prospects for whale shark conservation in Eastern Indonesia through Bajo traditional ecological knowledge and community-based monitoring. *Conservation and Society*, **10**(1):63–83.

Kaushik, H. (2013). "Fisherman rescues whale shark pup." *Times of India*, May 7, 2013, <articles.times ofindia.indiatimes.com/2013-05-07/ahmedabad/39089310_1_whale-sharks-pup-fishing-net>, accessed May 20, 2013.

Lederach, J. (2003). *The Little Book of Conflict Transformation: Clear Articulation of the Guiding Principles by a Pioneer in the Field*. Intercourse, PA: Good Books.

Lilette, V. (2006). Mixed results: conservation of the marine turtle and the red-railed tropicbird by Vezo semi-nomadic fishers. *Conservation and Society*, **4**(2):262–86.

Madden, F. (2004). Can traditions of tolerance help minimise conflict? An exploration of cultural factors supporting human–wildlife coexistence. *Policy Matters*, **13**:234–41.

Madden, F. and McQuinn, B. (2014). Conservation's blind spot: the case for conflict transformation in wildlife conservation. *Biological Conservation*, **178**:97–106.

Magannon, E. T. (1998). Where the spirits roam. *UNESCO Courier*, **51**(7/8):62–3.

Mallarach, J. (2011). Spiritual and religious values of northern Mediterranean wetlands: challenges and opportunities for conservation. In Papayannis, T. and Pritchard, D. E. (eds), *Culture and Wetlands in the Mediterranean: An Evolving Story*. Athens: Med-INA. pp. 347–61.

Mallarach, J. and Papayannis, T. (2010). Sacred natural sites in technologically developed countries: reflections from the experience of the Delos Initiative. In Verschuuren, B., Wild, R., McNeely, J. A. and Oviedo, G. (eds), *Sacred Natural Sites: Conserving Nature and Culture*. Gland: IUCN/Earthscan, pp. 198–208.

McClanahan, T. R., Glaesel, H., Rubens, J., and Kiambo, R. (1997). The effects of traditional fisheries management on fisheries yields and the coral-reef ecosystems of southern Kenya. *Environmental Conservation*, **24** (2):105–20.

Millennium Ecosystem Assessment. (2005). *Ecosystems and Human Well-Being: Synthesis*. Washington, DC: Island Press.

Murali Sankar, K. H. (2014). "Slaughter of whale sharks on the rise." *The Hindu*, June 23, 2014, <http://m.thehindu.com/news/national/andhra-pradesh/slaughter-of-whale-sharks-on-the-rise/article6139439.ece/>, accessed July 12, 2014.

Narayan, D., Patel, R., Schafft, K., Rademacher, A., and Koch-Schulte, S. (1999). *Can Anyone Hear Us? Voices from 47 Countries. Voices of the Poor*, Vol. 1. Washington, DC: World Bank.

Nichols, W. J. and Palmer, J. L. (2006). *The Turtle Thief, the Fishermen and the Saint: A Report on the Consumption of Sea Turtles during Lent*. Frankfurt: WWF.

Oppili, P. (2013). "Caught and dragged to death: a whale shark tale." *The Hindu*, January 29, 2013, <http://www.thehindu.com/news/cities/chennai/caught-and-dragged-to-death-a-whale-shark-tale/article4354641.ece>, accessed May 20, 2013.

Palmer, M. and Finlay, V. (2003). *Faith in Conservation: New Approaches to Religions and the Environment*. Washington, DC: The World Bank.

Pew Forum on Religion and Public Life. (2012). *The Global Religious Landscape: A Report on the Size and Distribution of the World's Major Religious Groups as of 2010*. <http://www.pewforum.org/uploadedFiles/Topics/Religious_Affiliation/globalReligion-full.pdf>, accessed July 13, 2013.

Pravin, P. (2000). Whale shark in the Indian coast—need for conservation. *Current Science*, **79**(3):310–15.

Rutte, C. (2011). The sacred commons: conflicts and solutions of resource management in sacred natural sites. *Biological Conservation*, **144**(10):2387–94.

Shepherd, S. and Terry, A. (2004). The role of indigenous communities in natural resource management: the Bajau of the Tukangbesi Archipelago, Indonesia. *Geography*, **89**(3):204–13.

Shores of Silence. (2001). [Film] Directed by Mike Pandey. New Delhi: Riverbank Studios.

Stacey, N. (2007). *Boats to Burn: Bajo Fishing Activity in the Australian Fishing Zone. Asia-Pacific Environment Monograph 2.* Canberra: ANU E Press.

Sudhi, K. S. (2012). "Whale shark killing continues in Kerala." *The Hindu*, December 28, 2012, <http://www.thehindu.com/todays-paper/tp-national/tp-kerala/whale-shark-killing-continues-inkerala/article4247779.ece>, accessed May 20, 2013.

Taylor, S. M. (2007). What if religions had ecologies? The case for reinhabiting religious studies. *Journal for the Study of Religion, Nature and Culture*, **1**(1):129–38.

Times of India. (2013). "105 whale sharks rescued in the last two years along Gujarat coast." *Times of India*, March 23, 2013, <http://articles.timesofindia.indiatimes.com/2013-2003-23/the-good-earth/37959674_1_whale-shark-gujarat-coast-rescue-teams>, accessed July 20, 2013.

UNEP–World Conservation Monitoring Centre. (2013). *UNEP-WCMC Species Database: CITES-Listed Species*, <http://www.cites.org/eng/resources/species.html>, accessed August 2, 2013.

Vincent, A. (2011). "El Santo Niño of the coral reef." Project Seahorse, June 27, 2011, <http://seahorse.fisheries.ubc.ca/node/403>, accessed August 1, 2011.

White, L. (1967). The historical roots of our ecologic crisis. *Science*, **155**(3767):1203–7.

Wildlife Trust of India. (2011). *Whale Shark Conservation Project.* <http://www.wti.org.in/oldsite/project-in-focus/july2010-whale-shark-conservation-project.html#updates>, accessed June 30, 2013.

Conclusion

Megan M. Draheim, Julie-Beth McCarthy, E. C. M. Parsons, and Francine Madden

This book looked at a diverse set of taxonomic, geographic, cultural, and political scenarios to help inform the conversation about human–wildlife conflict. It did not, however, contain a chapter on what might be considered "traditional" human–wildlife conflict (such as sharks biting people). Although this was not by design, it serves to demonstrate that human–wildlife conflict is rarely exclusively about direct conflict between a person and an animal (even when that direct conflict exists). Rather, human–wildlife conflict involves complex, multifaceted relationships and the shared histories people have with each other. Consequently, technical solutions are rarely enough to fully reconcile the conflict, and care must also be taken to first analyze the conflict fully and then thoughtfully and deliberately design the processes by which the parties can communicate and reach agreement. This is true even when wildlife is only marginally involved. Chapter 2 is a good reminder that conflict over places and ecosystems is just as important to address from a conservation perspective as the more commonly thought-of conflict that involves a specific species.

The levels of conflict model discussed throughout this book is one tool that can be used to aid conservation practitioners and researchers in their conflict analyses, a necessary first step to conflict reconciliation. Once these basic tools are learned, it is just as important for conservation professionals to keep these skills honed with practice and further study as it is for them to keep up with the latest technologies and best practices for natural science research.

Many themes come up repeatedly throughout the book, but this conclusion will focus on the importance of three: process, the social sciences, and the need for heightened awareness and sensitivity when "outsider" status comes into play.

Process

A theme that came up repeatedly throughout the book is that process matters. Whether it is a decision-making process or an attempt to transform conflict, if the process is not carefully considered and created, progress will be difficult to make. Francine Madden and Brian

McQuinn dove into this in Chapter 1, with an overview of why process matters and where it fits into the bigger picture of human–wildlife conflict.

Booker and Maycock's two case studies in the Bahamas (Chapter 2) provided insight into a participatory process on a local level, where one example worked and one did not. While many discussions of projects describe what worked, such candid discussions of projects and initiatives that did not go well can provide as much—or even more—help to the conservation community. In their examples, the process that included all of the important stakeholders in the issue (in this case, keeping the spiny lobster fishery sustainable and viable) early on was successful, while the process that neglected to include an important stakeholder (yachters) was not. An emphasis on relationship-building and ownership of both the problem and the solution in the spiny lobster initiative led to a result that satisfied all parties.

Jill Lewandowski and Carlie Wiener's chapters (Chapters 3 and 8, respectively), along with the chapter by Rachel Sprague and Megan Draheim (Chapter 7), tackled the process issue from a national government level. All three chapters described how the policy decision-making process in the U.S. federal government can serve to exacerbate conflict as a result of its design. Sprague and Draheim described how some Native Hawaiians, and others concerned with the presence of monk seals on the Main Hawaiian Islands (and the resultant regulations), are dissatisfied with the process because they feel their voice is not well heard. Wiener detailed how operators of swim-with-dolphin boat trips are dissatisfied with the regulatory decision-making process and how this has increased conflict over time. Lewandowski expanded on this, noting that the decision-making process can be problematic because of its linear structure, and its subsequent oversimplification and fracturing of the overarching issue into smaller pieces that each then need to be addressed separately— instead of holistically. It can also be problematic because of the pressure that timelines can produce: an overemphasis on science to the neglect of the social, political, and cultural issues that also play a role; the feeling that stakeholders are not part of the process; and the way that the current process does not address the conflict itself. Lewandowski then offered strategies that the government could use to transform their decision-making processes to address the deeper issues at play while staying within the regulatory guidelines that define their mandate. Although Chapters 3, 7 and 8 were focused on the United States, comparable regulatory and policy procedures are found in other parts of the world, so similar lessons can be drawn.

Christine Gleason also tackled the policy process from a national level, this time in the Dominican Republic, where whale-watching regulations for Santa Barbara de Samaná are created in the capital city of Santo Domingo (see Chapter 4). Here, there is a basic inequity issue which makes it difficult for all those but the owners of large-boat whale-watching operations to have their voices heard in the regulatory process. This leads to many important perspectives being ignored and important stakeholders being excluded, thus creating a situation ripe for conflict. Staying in the Caribbean region, Sarah Wise looked at MPA creation in the Bahamas and how the proposal to create a new MPA on Andros Island was seen by many locals as a way to advance the cause of a small group of elites (Chapter 6). While community meetings were held, they were often poorly attended and very tense, as distrust was high. In addition, many participants in these meetings wanted to focus on already established MPAs, as they felt that the process that had created these parks had not been satisfactory. Since there was no process to address these deeper conflicts, there is ongoing conflict over the latest MPA.

E. C. M. Parsons came at this issue from an international perspective, with his discussion of the IWC and the ongoing conflict between anti-whaling and pro-whaling countries (Chapter 5). He argued that the treaty commission is set up in a way that focuses on the question of science (e.g., does Japan's scientific whaling program provide valid and valuable data) and takes away from the deeper questions of cultural identity and norms that might be driving the conflict. For example, there is a basic mismatch between how pro- and anti-whaling countries perceive whales. Japan thinks of them as "just another fish," a resource to be harvested, while many Western countries grant whales a special status that elevates concern for them. The current regulatory process, however, does not address the cultural and social differences between the various sides of the issue (this also includes charges of racism and colonialism).

Finally, Julie-Beth McCarthy's chapter (Chapter 9) tackled the role of technical solutions in a conflict transformation process. In Gujarat, India, a compensation program for fishermen to release whale sharks was set up after an extensive campaign that worked to transform the image of the whale shark from a resource to a symbol of cultural, religious, and geographic pride. As McCarthy noted, the technical solution (in this case, the compensation program) was arguably only successful because it came after the initiative that sought to change Gujarat residents' opinions of the whale shark. Although technical solutions can provide meaningful conservation help, it might be more effective for the technical solutions to be addressed after the more difficult social, political, and cultural issues have been addressed. In other words, the more complex, human-centric issues should be addressed in the process before technical solutions are implemented.

The social sciences

While the conservation community has come a long way toward embracing the social science world and their methodologies, this acceptance is not yet universal. Many of the chapters in this book demonstrated the importance of bringing the social sciences into attempts to transform deep underlying and identity-based conservation and human-wildlife conflicts. Several chapters (including Chapters 3, 5, 7, and 9) demonstrated that natural science knowledge is not enough to tackle these deeper issues.

While there is a tendency in the conservation community to equate the social sciences with quantitative, questionnaire-based research, this book also highlights the importance of qualitative research. Lewandowski, Gleason, Wiener, and Wise all drew directly from their own qualitative research, while Booker, Draheim, Madden, Maycock, McCarthy, McQuinn, Sprague, and Parsons drew from social science theory (much of which has been established using qualitative methodology) as well as informal qualitative methods such as informal interviews and observations. Qualitative research methodology enables researchers and practitioners to gain a deeper understanding of what is happening in the hearts and minds of parties to a conflict, whereas quantitative research methodology (although it does certainly provide valuable information) tends to focus on whether what the researcher thinks is occurring is actually happening. In a survey, the researcher is the one who writes the questions and provides the potential answers. If he or she misses an important piece of the puzzle, there is no reliable way to account for that missed information. Studies that use qualitative or a mix of qualitative and quantitative methodologies might provide richer data. Related to this is the creation of process design, discussed in more detail in the previous section ("Process"). The social sciences are the underpinnings of good process design; this observation was highlighted most directly in the first three chapters, although it was discussed throughout the book.

The importance of social sciences also speaks to the value of developing a conservation-related skill set that not only includes technical skills such as GIS and effective sampling strategies but also includes "softer" skills such as the ability to conduct constructive interviews and focus groups, facilitate meetings, and (again) design processes that address conflict in a productive way. In particular, Chapter 1 addressed the need to not only gain these skills but also practice them over time.

Outsider status

The final theme that will be discussed here is the role a group or individual labeled as an outsider can play in conflict. This manifests itself in many ways. For example, Chapters 4, 6, and 7 all described situations where advocates and researchers who are not local and thus may not be as sensitive to local conflict dynamics can be looked upon with suspicion (or worse) by locals who are affected by the conservation issues at play. Gaining the knowledge and skills that will help conservation practitioners and researchers navigate this role while participating in a constructive manner is important. Chapter 1 provided some guidance as to how wildlife professionals can achieve this.

This outsider status can also be linked to tourism, where either the tourists themselves are seen as the outsiders (e.g., in Chapter 2) or the tourism operators (e.g., some of the operators in Chapters 4, 6, and 8) are seen as nonlocal. This perception can cause complex problems, especially in areas where tourism is a big (if not the biggest) contributor to the economy.

It is not uncommon for an identity conflict that started generations ago to impact the local perceptions of and receptivity to someone with "outsider" status, even if the outsider had nothing to do with the creation or perpetuation of that conflict. For instance, several chapters noted that conservation efforts today are impacted by a history of colonialism: for example, the chapters set in the Bahamas (Chapters 2 and 6), Hawai'i (Chapters 7 and 8), the Dominican Republic (Chapter 4), and insular Southeast Asia (Chapter 9). In Chapter 5, Parsons also tackled colonialism, both directly (Japan's allies at the IWC are mostly countries that were colonized by Western countries at some point in their history) and more indirectly through charges of cultural colonialism, where anti-whaling nations are seen as attempting to force their views of whales on Japan and other whaling nations. The shadows these colonial histories cast can be hard to notice in a room full of brightly lit disputes, but they can be directing the conflict in subtle ways. To that end, it is critical that individuals and groups with "outsider" status develop a greater awareness of and sensitivity to the spectrum and history of conflicts that exist within the social system in which conservation takes place, as well as recognize both the opportunities and challenges present in this role.

Conclusion

This book has covered a lot of ground, both taxonomically and geographically. It serves as an introduction to this complex field for conservation practitioners, researchers, and students as well. We hope our readers will see that despite this complexity, solutions to marine human–wildlife conflict exist that help to conserve marine species while at the same time supporting human communities. Solutions that do both offer our best chance at reaching our conservation goals.

Index

Page numbers followed by 'f' indicates entries in figures, 't' in tables and 'b' in boxed material.

UNIVERSITIES AT MEDWAY LIBRARY